REACTION KINETICS IN MICELLES

REACTION KINETICS IN MICELLES

Proceedings of the American Chemical Society Symposium on Reaction Kinetics in Micelles, New York, New York, August 1972

Edited by EUGENE CORDES
Department of Chemistry
Indiana University
Bloomington, Indiana

PLENUM PRESS • NEW YORK–LONDON • 1973

Library of Congress Catalog Card Number 72-94823
ISBN 0-306-30722-7

© 1973 Plenum Press, New York
A Division of Plenum Publishing Corporation
227 West 17th Street, New York, N.Y. 10011

United Kingdom edition published by Plenum Press, London
A Division of Plenum Publishing Company, Ltd.
Davis House (4th Floor), 8 Scrubs Lane, Harlesden, London, NW10 6SE, England

All rights reserved

No part of this publication may be reproduced in any form without
written permission from the publisher

Printed in the United States of America

PREFACE

This volume is a collection of the presentations given at a symposium on reaction kinetics in micelles at the 164th Meeting of the American Chemical Society, Division of Colloid and Surface Science, in New York City in August of 1972.

Participants in this symposium have all been active in the field of reaction kinetics in micelles during the past several years. Indeed, much of what is known in this area is the direct result of their efforts. It is hoped that this small volume will serve to summarize the current state of knowledge in this area, to point out some new directions toward which research efforts are pointed, and to induce new researchers with fresh points of view to enter this young and relatively unexplored area of chemistry.

CONTENTS

List of Authors.................................... ix

Recent Advances in the Chemistry of Micelles......... 1
    Norbert Muller

Micellar Catalysis for Carbonium Ion Reactions....... 25
    Jan Baumrucker, Maritza Calzadilla, and
    E. H. Cordes

Micellar Effects in Steady-State Radiation Induced
    Reactions....................................... 53
    J. H. Fendler, G. W. Bogan, E. J. Fendler,
    G. A. Infante, and P. Jirathana

Electrolyte Effects on Micellar Catalysis............ 73
    C. A. Bunton

Micellar Control of the Nitrous Acid Deamination
    Reaction.. 99
    Robert A. Moss, Charles J. Talkowski, David W.
    Reger, and Warren L. Sunshine

Catalysis by Inverse Micelles in Non-Polar Solvents.. 127
    E. J. Fendler, Shuya A. Chang, J. H. Fendler,
    R. T. Medary, O. A. El Seoud, and V. A. Woods

Author Index 147

Subject Index 151

LIST OF AUTHORS

JAN BAUMRUCKER
   Escuela de Quimica, Universidad Central
   Caracas, Venezuela

G. W. BOGAN
   Department of Chemistry, Texas A & M University
   College Station, Texas 77843

C. A. BUNTON
   Department of Chemistry, University of California
   Santa Barbara, California 93106

MARITZA CALZADILLA
   Escuela de Quimica, Universidad Central
   Caracas, Venezuela

SHUYA A. CHANG
   Department of Chemistry, Texas A & M University
   College Station, Texas 77843

EUGENE H. CORDES
   Department of Chemistry, Indiana University
   Bloomington, Indiana 47401

O. A. EL SEOUD
   Department of Chemistry, Texas A & M University
   College Station, Texas 77843

E. J. FENDLER
   Department of Chemistry, Texas A & M University
   College Station, Texas 77843

J. H. FENDLER
   Department of Chemistry, Texas A & M University
   College Station, Texas 77843

G. A. INFANTE
   Department of Chemistry, Texas A & M University
   College Station, Texas 77843

LIST OF AUTHORS

P. JIRATHANA
    Department of Chemistry, Texas A & M University
    College Station, Texas 77843

R. T. MEDARY
    Department of Chemistry, Texas A & M University
    College Station, Texas 77843

ROBERT A. MOSS
    Wright Laboratory, School of Chemistry, Rutgers
    University, The State University of New Jersey,
    New Brunswick, New Jersey 08903

NORBERT MULLER
    Department of Chemistry, Purdue University
    Lafayette, Indiana 47907

DAVID W. REGER
    Wright Laboratory, School of Chemistry, Rutgers
    University, The State University of New Jersey,
    New Brunswick, New Jersey 08903

WARREN L. SUNSHINE
    Wright Laboratory, School of Chemistry, Rutgers
    University, The State University of New Jersey,
    New Brunswick, New Jersey 08903

CHARLES J. TALKOWSKI
    Wright Laboratory, School of Chemistry, Rutgers
    University, The State University of New Jersey
    New Brunswick, New Jersey 08903

V. A. WOODS
    Department of Chemistry, Texas A & M University
    College Station, Texas 77843

RECENT ADVANCES IN THE CHEMISTRY OF MICELLES

Norbert Muller

Department of Chemistry, Purdue University

Lafayette, Indiana 47907

INTRODUCTION

During the last decade the study of micelle formation in detergent solutions, like many other areas of chemical science, has undergone almost explosive growth, promoted by technological advances which made possible substantial improvements in existing experimental methods and the introduction of altogether new techniques. An additional stimulus has been an upsurge of interest in the micellization phenomenon among chemists not primarily concerned with surfactant solutions for their own sake. This group consists principally of biochemists who have focussed attention on similarities between detergent micelles, monolayers, or bilayers and the phospholipid regions of biological membranes, and between micelles and globular proteins. Although micelles and proteins do not really have very much in common from a structural point of view beyond the fact that both are stabilized by hydrophobic interactions, it is now well established [1,2] that micelles exert catalytic effects on assorted organic reactions which are at least reminiscent of the effects produced by enzymes. This discovery aroused the interest of physical organic chemists, and its ramifications form the subject matter of the major portion of this symposium.

Fortunately many of the newer contributions in micelle chemistry are cited in a recent review [2]; space would not allow even a brief mention of all these publications here.

What is attempted here is not a comprehensive coverage of recent advances but a report on three problem areas, where new information has become available but it is not yet completely clear how it should be interpreted, together with several suggestions which are offered in the hope of helping to resolve some apparent contradictions. These areas are:
1. Kinetics of micelle formation, dissolution, and dissociation.
2. Location of water and of solubilized molecules in micelles.
3. Changes in thermodynamic variables for micellization.

KINETICS OF MICELLE FORMATION, DISSOLUTION, AND DISSOCIATION

Above the critical micelle concentration (cmc) a surfactant solution contains monomers (S) and an assortment of aggregates varying in size and stability. This not only means that a large number of rate or equilibrium constants is required to characterize the system completely, but it also introduces some semantic problems. In particular, the phrase "micelle dissociation" has been applied to either of the two reactions

$$S_n \underset{k_{n-1,n}}{\overset{k_{n,n-1}}{\rightleftarrows}} S_{n-1} + S \tag{1}$$

and

$$S_n \rightleftarrows nS, \tag{2}$$

where S_n is the most abundant micellar species, with aggregation number n. Equation (2) of course represents the overall result of sequence of n-1 similar reactions of which (1) is the first, and before one can discuss a rate of micelle dissociation it must be made clear which process is meant. "Micelle dissociation" will be used here only to denote process (1), and (2) will be called "micelle dissolution." For an ionic micelle which binds counterions (X) or one which contains a solubilized molecule (G) two additional dissociation processes can be written,

$$S_n X_m \rightleftarrows S_n X_{m-1} + X \tag{3}$$

and

$$S_nG \underset{k_+}{\overset{k_-}{\rightleftarrows}} S_n + G . \qquad (4)$$

These may be called "micelle-counterion dissociation" and "micelle-guest dissociation".

It has long been recognized that in fairly dilute aqueous solutions all these equilibria involve half-lives which are short compared with the time usually required to mix the components [3,4]. Recently several of the new techniques for the study of fast reactions have been used in attempts to measure the rate constants $k_{n,n-1}$ or k_- of reaction (1) or (4). Unfortunately, these studies have produced apparently incompatible results.

The experimental approach with which the author is most familiar, i. e. nmr spectroscopy, has so far yielded data which make it possible only to set a lower limit on $k_{n,n-1}$ or k_-. This requires the use of a species S or G with at least one nmr signal which changes position when the molecule is transferred from the bulk solution to the micelles. Exchange of material between monomeric and micellar sites is always found to be fast enough to cause a single time-averaged spectrum to appear. For the selected signal, the chemical shift expressed in frequency units is

$$\langle \nu \rangle = f_{mic} \nu_{mic} + (1 - f_{mic}) \nu_{aq} , \qquad (5)$$

where f_{mic} is the fraction of the material in the micellized form, and the two limiting shifts ν_{mic} and ν_{aq} can be determined graphically from the variation of $\langle \nu \rangle$ with concentration [5]. Standard nmr theory [6] gives for the width of this resonance at half height

$$\pi \Delta \nu = (1/T_2)_{eff} = f_{mic}(1/T_2)_{mic} + (1 - f_{mic})(1/T_2)_{aq} + (1/T_2)_{exch} . \qquad (6)$$

The terms $(1/T_2)_{aq}$ and $(1/T_2)_{mic}$ are the transverse relaxation times in water and in the micelle, which are approximately equal and of the order of 1 sec^{-1}. The term $(1/T_2)_{exch}$ represents the additional line broadening which is expected unless the exchange process is very fast and is related to the reaction rate through

$$(1/T_2)_{exch} = 4\pi^2 f_{mic}^2 (1-f_{mic})^2 (\nu_{mic}-\nu_{aq})^2 (\tau_{mic}+\tau_{aq}) , \quad (7)$$

where τ_{mic} and τ_{aq} are mean residence times for S or G in the micelles and in the bulk solution. In nmr spectra of a wide variety of micellar solutions [5,7-12] <u>such additional broadening has never been reported</u>. With the reasonable assumption that a change of 1 Hz in the line width would have been noticeable, this gives an upper limit of about 3 sec^{-1} for $(1/T_2)_{exch}$ for all systems studied.

The further implications of this result depend to some extent on the material chosen, the type of dissociation involved, and the operating conditions. When the exchange process is (1) it is easy to select data for a concentration where $f_{mic} = 1/2$; then $\tau_{mic} = \tau_{aq}$ and (7) reduces to

$$(1/T_2)_{exch} = (\pi^2/2)(\nu_{mic}-\nu_{aq})^2 \tau_{mic} , \quad (8)$$

and the absence of detectable broadening implies

$$\tau_{mic} \leq \frac{6 \text{ sec}^{-1}}{\pi^2 (\nu_{mic}-\nu_{aq})^2} . \quad (9)$$

In 60 MHz proton spectra the peaks having the largest shifts usually arise from phenyl groups [7,9] and typically $\nu_{mic}-\nu_{aq}$ is about 15 Hz; then $\tau_{mic} \leq 3 \times 10^{-3}$ sec. Fluorine magnetic resonance spectroscopy is attractive primarily because solvent effects on the chemical shifts are larger, and for a variety of anionic, cationic, and nonionic detergents having fluorine labelled alkyl chains $CF_3(CH_2)_n$- with $7 \leq n \leq 12$ we have found [5,10-12] $\nu_{mic}-\nu_{aq}$ = 66Hz with a spectrometer frequency of 56.4 MHz, giving $\tau_{mic} \leq 1.4 \times 10^{-4}$ sec. A proton nmr study of benzene solubilized in sodium dodecyl sulfate micelles [7] likewise indicated that in a reaction of type (4) $\tau_{mic} \leq 10^{-4}$ sec.

Although τ_{mic} probably depends somewhat on the size of the molecules, it should not be drastically affected by introduction of the fluorine atoms, since their effect on the stability of micelles has been shown [11] to be quite small. It thus seems likely that τ_{mic} is uniformly less than 1.4×10^{-4} sec, at least for small guest molecules and for detergents having 13 carbon atoms or less in the alkyl chain and headgroups no larger than $(OC_2H_4)_6OH$.

Turning to the rate constants, the simplest case is reaction (4) where $k_- = 1/\tau_{mic}$ and therefore must exceed 10^4 sec^{-1} for benzene in dodecyl sulfate micelles. For process (1) allowance must be made for the fact that on the average the reaction must proceed n times in each direction before a given monomer unit is released from the micelle, whence

$$k_{n,n-1} = n/\tau_{mic} . \qquad (10)$$

As a rule the aggregation number is 50 or larger, which then requires $k_{n,n-1} \geq 3 \times 10^5$ sec^{-1}; a few nonionic detergents form substantially smaller micelles and thus could have smaller dissociation rate constants.

These conclusions are supported by studies of ultrasonic absorption in dilute micellar solutions of sodium dodecyl sulfate and a number of alkyl carboxylates [13-15]. The process giving rise to the relaxation frequency near 0.1 MHz for sodium dodecyl sulfate was first believed to be micelle-counterion dissociation, but this appears improbable in view of the later work with a variety of soaps, which showed that the frequencies were strongly dependent on the alkyl chain length and almost completely unaffected by changing the counterions from sodium to potassium to cesium. The relaxation frequencies were somewhat concentration dependent and ranged from about 0.3 MHz for potassium dodecanoate to about 25 MHz for the heptanoate. They were interpreted with reference to a proposed mean dissociation reaction intermediate between (1) and (2), namely

$$S_{n+m} \underset{k_1}{\overset{k_{-1}}{\rightleftharpoons}} S_{n-m} + 2mS , \qquad (11)$$

where n is the most probable aggregation number and m increases from about 4 to 10 with increasing alkyl chain length. Reported values of the rate constants k_{-1} included 1.8×10^5 sec^{-1} for the dodecanoate, increasing up to 4.1×10^7 sec^{-1} for the heptanoate.

Very recently the rate of micelle-guest dissociation has been measured by electron spin resonance [16] for di-t-butyl nitroxide in solutions of sodium dodecyl sulfate. Depending on the relative concentrations of the spin-label and the detergent the ratio of the concentration of nitroxide in micelles to that in water varied between about 0.5 and 2.0. The free and micelle-bound species produced separate spectra, showing that process (4) was slow on the ESR time scale. At higher temperatures the exchange rate increased sufficiently to cause measurable changes in the line shapes from which k_- could be evaluated. Using the Arrhenius equation to extrapolate these results to 25° one obtains $k_- = 4.3 \times 10^5$ sec^{-1}, quite in accord with the implications of the nmr work.

In sharp contrast with the above are the results of a series of studies in which detergent solutions were perturbed by a temperature jump, and either absorption spectroscopy or light scattering photometry was used to follow the concentration of micellized material as a function of time [17-20]. Relaxation times ranged between about 1 and 300 msec, depending on the nature of the headgroup, the alkyl chain length, and the concentration. With the assumption that (1) is the rate-determining step for the observed relaxation, the following values of $k_{n,n-1}$ were derived:

4 sec^{-1} for dodecyl ammonium chloride [19],
17 sec^{-1} for dodecyl sodium sulfonate [19],
120 sec^{-1} and 50 sec^{-1} for dodecyl pyridinium iodide [17,19],
5.3 sec^{-1} for sodium dodecyl sulfate [18],
73 sec^{-1} and 0.42 sec^{-1} for OPE30 and OPE16 [20].

The last two materials are nonionic octylphenyl polyoxyethylene ether detergents. It is very disturbing that $k_{n,n-1}$ in these systems would appear to be several orders of magnitude smaller than it is for the many rather similar materials studied by nmr. The only available rate constant based on a stopped-flow study [21] is larger than those deduced from T-jump data but still much below the lower limit suggested by the nmr observations. This, however, may be due to the fact that the detergent involved has a $(OC_2H_4)_{30}OH$ headgroup, much bulkier than those of any materials used in the fluorine nmr studies.

The crucial difference between the exchange process which affects nmr and ESR line widths and the relaxation which follows a T-jump or a sudden dilution is that micelle dissociation (1) alone suffices to bring about chemical exchange, but the T-jump or stopped flow methods follow changes in micelle concentrations which require dissolution (2) or its reverse. Originally the T-jump data were interpreted on the assumption that when (2) proceeds in either direction (1) is the only slow step and therefore determines the overall rate. The work cited here in effect provides an experimental test of this mechanism and suggests that it ought to be discarded.

A mathematically acceptable way of reconciling the various results is to assume [21] that the slow step in micelle dissolution is not (1) but a subsequent reaction

$$S_{n-j} \rightleftarrows S_{n-j-1} + S . \tag{12}$$

This mechanism requires that the species S_{n-j} be several orders of magnitude longer-lived than the principal micellar species, although by definition S_n is more stable than S_{n-j}. This is intuitively very difficult to accept. The author has recently suggested [22] an alternative model in which micelle dissolution involves a sequence of n-2 steps of equal rate constant, the last step then being the rapid dissociation of the dimer. During the T-jump experiment, the state of the system is at all times quite close to the equilibrium state, and it is proposed that each change in size for a particular micelle has about an equal chance of being an increase or a decrease. Then the dissolution process becomes analogous to a one-dimensional random walk with the final state lying n-2 steps away from the initial state and a time interval between steps of the order $\Delta t \simeq 1/2k_{n,n-1}$. On the average, the net distance covered in a random walk increases with the square root of the number of steps taken, so that the mean time required for dissolution of a micelle is approximately

$$\tau \simeq \frac{n^2}{2k_{n,n-1}} . \tag{13}$$

The analysis leading to (13) is quite obviously oversimplified. There is no compelling reason to suppose that all rate constants in such a multiple-step mechanism should be exactly equal, or that the number of slow steps is n-2 rather than a somewhat smaller number. The equation provides no rationale for a linear dependence of the relaxation time on the total detergent concentration, regularly found with the T-jump method. The existence of counterions for the ionic detergents is entirely ignored, but this was done also in the earlier analysis [17]. In spite of all these caveats, the treatment offers the one great advantage that with plausible assumptions it yields an equation (13) which makes relaxation times in the msec range compatible with $k_{n,n-1}$ values of 10^5 to 10^7 sec^{-1}, when n is about 50 or larger. It would appear that the many-step model corresponds more nearly to physical reality than a single-step mechanism and is worthy of further attention.

Estimated rate constants obtained with (13) suggest that exchange effects might be observable in fluorine nmr spectra if a superconducting magnet with a field strength of 52 kgauss were used so that the term $\nu_{mic}-\nu_{aq}$ in equation (7) would become about 240 Hz. Such studies would then provide a simple direct way of exploring the effects of chain length, concentration, temperature, or other variables on micellization rates.

LOCATION OF WATER AND OF SOLUBILIZED MOLECULES IN MICELLES

It is generally agreed that micelles are hydrated species. For detergents containing octyl or dodecyl chains, hydrodynamic data show that as many as 10 to 12 water molecules are bound per detergent ion [23]. The location of these water molecules presents a problem for which no unequivocal answer is yet available. As discussed below, any attempt to determine the position of a solubilized organic molecule within a micelle requires some assumption about the arrangement of the water molecules, so that it becomes expedient to treat these two problems together.

Any reference to the location of a component of a micelle must not be taken to imply that the micelle is a rigid entity with permanent structural properties. Over the years it has often been said that micelles are more like small liquid drops than like crystallites, first because this was a plausible assumption, then because of a growing body of evidence [4]. Recent results of ESR spectroscopy [24,25] and fluorescence depolarization measurements [26] dramatically confirm the fluidity of micelles. The basis for the ESR approach is that in the spectrum of a paramagnetic probe molecule the line width is determined primarily by the rate of rotation, as shown by the effects of changing the solvent viscosity. Broad lines are obtained not only in very viscous media but also when a small probe molecule is bound to a rigid macromolecule, which slows its tumbling rate by several orders of magnitude. The sharp ESR spectra obtained with solubilized nitroxide derivatives show that the molecules rotate almost as rapidly when micelle-bound as they do when dissolved in water alone.

The results with fluorescent dyes are rather similar. The essential idea is that when polarized light is used to excite the fluorescence, the extent of depolarization of the emitted light depends on the average angle through which the molecule can rotate during the very short time interval between absorption of the incident photon and emission of the fluorescence. Like the ESR line width, the degree of depolarization is therefore dependent on the solvent viscosity, or on the local "microviscosity" which in a micellar solution may differ from the overall value. With perylene and 2-methylanthracene in micelles of several cationic detergents, microviscosities in the range 17 to 50 cP were reported [26]. For comparison, the viscosity of liquid oleic acid at 30° is 25.6 cP.

Even though these studies show that micelles are not at all similar to rigid polymers with fixed binding sites for small molecules, many observations indicate that different guest molecules tend on the average to be found in different regions of the micelles, some near the periphery and others close to the center. This is most often shown with the help of a probe molecule having some measurable property P with a value that changes appreciably, from P_{aq} to P_{hc}, when the material is transferred from water to a hydrocarbon solvent. P might be a proton or fluorine chemical shift, a g-value or coupling constant for a paramagnetic species, or the wavelength of some convenient electronic transition. When the compound is dissolved in a detergent solution, the value of P in the micellar environment can be observed directly either if exchange between micellar and extramicellar sites is slower than some characteristic rate depending on the experiment used or if the material is so insoluble in water that it is safe to assume that it is present in solution only as the micelle-guest complex. Otherwise, an average value will be found obeying an equation similar to (5), i. e.

$$<P> = f_{mic}P_{mic} + (1-f_{mic})P_{aq} , \qquad (14)$$

and the determination of P_{mic} then requires supplementary experiments which provide a reliable estimate of f_{mic}. Once P_{mic} is at hand, one may define a parameter [5]

$$Z = (P_{mic} - P_{aq})/(P_{hc} - P_{aq}) , \qquad (15)$$

which is usually found to have a value between 0 and 1. When $Z \sim 0$, it is presumed that the guest molecule is in a very water-like environment and that it is bound at the micelle surface. When $Z \sim 1$, the environment is hydrocarbon-like and it is inferred that the guest is buried in the micellar interior. Often intermediate values of Z are found [5,27] even for species as hydrophobic as benzene and benzotrifluoride, suggesting that the probes are only partially removed from contact with the aqueous medium.

One difficulty with this procedure is that P_{mic} and Z depend on the concentration of the detergent and the solubilized material [5,8,27-30]. Such a dependence can arise in at least three ways. Increasing amounts of solubilizate can cause swelling of the micelles and perhaps pronounced changes in their internal organization. Solubilized molecules in the same micelle may eventually affect each other, so that the environment of each contains species which may be quite unlike either water or saturated hydrocarbon. Finally, certain regions of the micelle may be preferred by the first few guest molecules, with less favorable sites coming into use only when the initial sites are saturated. As a result, values of P_{mic} obtained at different guest: detergent concentration ratios may have contradictory implications [28,29]. When feasible, it is probably best to extrapolate P_{mic} to zero solubilizate concentration before evaluating Z.

A more serious difficulty in the use of Z is that one cannot interpret values near 0.5 without knowing to what extent water penetrates into the micelles. This is most simply illustrated by fluorine nmr studies of detergents with a terminal trifluoromethyl group in the alkyl chain, where the detergent itself serves as the probe [5,10-12]. The experiments consistently give Z values close to 0.5 regardless of whether the micelles are anionic, cationic, or nonionic. This seems to rule out the possibility that the shifts are seriously affected by possible electrostatic interactions between the trifluoromethyl groups and the headgroups, so that the results may be taken as an indication that the trifluoromethyl groups are in a partially aqueous environment. On purely geometric grounds it is necessary for some of these groups to be near the micelle surface, but it is not possible to say unambiguously whether or not trifluoromethyl-water contacts occur only in the outermost region of the micelles.

Although this problem cannot be resolved at this time, some clarification may be obtained by attempting to formulate more precisely what is meant by the outer region of the micelle. The accepted model [2] for ionic micelles consists of a core with a radius about equal to the length of the fully extended alkyl chain of the detergent, surrounded by the Stern layer which contains water, the headgroups, and more than one-half of the counterions, and is in turn surrounded by the Gouy-Chapman double layer which contains water and the remaining counterions. It may be helpful to divide the core further, into an inner core from which water is excluded and an outer core into which the solvent may penetrate. The existence of such an outer core considerably simplifies the geometric problem of how to build an essentially spherical cluster out of linear detergent molecules, without leaving voids, while restricting the polar groups exclusively to surface sites. The fact that about 40% of the micellar volume consists of water which is bound firmly enough to move with the micelle during a diffusion or sedimentation process [23] also suggests that some water is in the outer core region, although other interpretations are possible. More nearly unequivocal supporting evidence comes from studies of the partial molal volume changes in micelle formation [31,32] and proton nmr spectra [33,34].

For dilute aqueous solutions of long-chain detergents, the partial molal volume in the monomeric form is always less than in the micellar state. This has been rationalized by supposing that water adjacent to a hydrocarbon chain has a smaller effective volume than bulk water. Then the process of micellization involves a volume increase because most of these loosely monomer-bound water molecules are set free. The volume difference decreases as the alkyl chain length decreases and extrapolates to zero for a hexyl chain. This suggests that several methylene groups adjacent to the solubilizing group are hydrated in nearly the same way in the micellar as in the monomeric state.

The nmr results include both chemical shifts and relaxation times. Sodium alkyl sulfates produced effects on the chemical shift of water protons [33] which differed depending on whether the detergents were present as monomers or as micelles. This difference again is chain-length dependent, and extrapolates to zero for a chain of about 3 or 4 carbon atoms, suggesting once more that the interaction between water and the first few methylene groups is not much altered by micellization. Relaxation times were measured [34] for the methylene protons of the same detergents, in D_2O solutions. For the monomers, the major relaxation mechanism involves interactions between protons in the same alkyl chain, and the relaxation time is in the range of 3 to 10 sec. In the micelles, interactions between protons of neighboring alkyl chains provide an additional relaxation process, and the relaxation times are reduced to between 0.3 and 1.5 sec. Again the effect is chain-length dependent. It appears that two or three methylene groups remain in a water-like (D_2O) environment even in the micelle, so that the fraction of the methylene protons actually in a hydrocarbon-like environment increases as the chain grows longer. It is also significant that the average rate of relaxation for methylene protons <u>in micelles</u> changes when H_2O replaces D_2O as the solvent.

Taken together, these findings imply that although no sharp boundary between inner and outer core can be defined, the outer core probably contains approximately the first four methylene groups. For a dodecyl sulfate micelle it would follow that the radius of the inner core is about 2/3 of the total core radius. A simple but almost startling consequence is that the inner core would make up only 8/27 or 30% of the core volume while the outer core volume is 70% of the total! Thus perhaps the reason why the Z parameter is seldom found to be much larger than 0.5 is that a guest molecule can escape completely from contact with water only by confining itself to a very small region near the micelle center. Any such restriction is of course opposed by thermal motions, i. e. is accompanied by a reduction in the entropy of the system.

A somewhat different and much less used technique for localizing a guest molecule is to observe perturbations of the spectrum of the host detergent caused by the solubilized species. For example, nmr signals from different protons of cetyl trimethylammonium bromide (CTAB) are unequally affected by solubilized benzene [8], especially when the benzene:CTAB mole ratio is in the range 0.1 to 1.0. The largest change is in the position of the α-methylene peak, suggesting that benzene is progressively replacing water in the outer core region. At higher benzene:CTAB ratios all the signals are increasingly affected, and the additional benzene apparently dissolves throughout the micelle.

THERMODYNAMICS OF MICELLIZATION

Since nearly all studies of dilute detergent solutions provide a value of the critical micelle concentration, and the standard free energy change of micellization per monomer is given to a good approximation [4] by

$$\Delta G_m^o = RT \ln cmc , \qquad (16)$$

many ΔG_m^o values have been accumulated over the years. Only in relatively recent times have there been systematic efforts to decompose ΔG_m^o into its enthalpic and entropic components,

$$\Delta G_m^o = \Delta H_m^o - T\Delta S_m^o . \qquad (17)$$

Values of ΔH_m^o have been determined calorimetrically for a number of detergents [35-37]. Other values have been calculated from the relation [4,37]

$$\Delta H_m^o = - RT^2 \frac{d}{dT} \ln cmc , \qquad (18)$$

which is applicable when the aggregation number and degree of ionization are temperature-independent. The agreement between the results of the two approaches is excellent for nonionic detergents and fair for ionic species. For many detergents ΔH_m^o is positive, and the favorable free energy of micellization must arise from a dominant $T\Delta S_m^o$ contribution. Even when ΔH_m^o is negative it is so much smaller than the total free energy change that one must still regard micellization as primarily an entropy directed process.

At first sight it seems very simple to rationalize this finding by a straightforward application of prevalent ideas on the nature of hydrophobic interactions. It is supposed that water in the immediate vicinity of a dissolved hydrocarbon chain is subject to enhanced hydrogen bonding and hence exists in a state of lower entropy and enthalpy than bulk water [38,39]. Consequently, transfer of an alkane molecule or an alkyl group from water to a nonpolar environment should involve a small, probably positive enthalpy change and a large, positive entropy change, just as one actually finds for the micellization process. Difficulties arise when one takes the next logical step towards a detailed understanding of micellization thermodynamics by dividing ΔH_m^o into a contribution from the hydrophilic group and a hydrocarbon chain contribution. This requires measurements of ΔG_m^o and ΔH_m^o as a function of chain length for homologous series of detergents.

It is well known [4] that for straight-chain detergents addition of one methylene group makes ΔG_m^o more negative by about 0.68 kcal/mol at 25°. Model theories of hydrophobic bonding suggest that this change is due primarily to an increase in the $T\Delta S_m^o$ term, not more than 20% of it being attributable to a more favorable ΔH_m^o. Indeed it has been stated [40] that the Nemethy-Scheraga theory requires ΔH_m^o to become more positive as the chain length increases, so that the increase in $T\Delta S_m^o$ must be even larger than the net change in ΔG_m^o. However, measured values of ΔH_m^o for several series of detergents, given in Table I, clearly show that

TABLE I

Chain-length dependence of the heat of micellization for homologous series of detergents. Values are given for ΔH_m^o in kcal per mole of material micellized at 25°.

| Headgroup | \multicolumn{5}{c}{Number of Carbon Atoms in Alkyl Chain} | Ref. |
	6	8	9	10	12	
$(OC_2H_4)_6OH$	5.5	4.8	---	3.6	---	36
$N(O)(CH_3)_2$	---	4.0	4.4	2.7	2.6	35
$OSO_3^-Na^+$	---	1.5	---	1.0	-0.3	35
$SO(CH_2)_2OH$	2.4	1.2	---	---	---	37
$SO(CH_2)_3OH$	2.7	1.7	---	---	---	37
$SO(CH_2)_4OH$	3.4	2.0	---	---	---	37

addition of a methylene group produces a change of about −0.6 kcal/mol in ΔH^o_m, accounting for almost 90% of the total change in ΔG^o_m and implying that the entropy of micellization increases only very slowly as the alkyl chain grows, at least in the range from hexyl to dodecyl. A closely similar trend has been noted [35,41] in the heats and free energies of solution of n-alkanols and n-alkyl hexaoxyethylene glycol monoethers, where each added methylene group contributes about 0.8 kcal/mol to the free energy of solution, of which more than two-thirds is enthalpic in origin.

In the light of these observations it is interesting to review the thermodynamic quantities for the transfer of the smallest alkanes from water to a hydrocarbon environment, collected in ref. 39 and given in Table II. The data support the following conclusions:

1. The overall negative values of ΔG^o_{tr} are indeed entropic rather than enthalpic in origin.

2. The average increase in $T \Delta S^o_{tr}$ per added methylene group is 0.47 kcal/mol, but extrapolation to the higher alkanes is very uncertain in view of the near equality of ΔS^o_{tr} for propane and butane.

3. The average increment in ΔH^o_{tr} per added methylene group is −0.6 kcal/mol, a trend identical with that found for the detergent ΔH^o_m values.

TABLE II

Enthalpy, entropy, and free energy changes for the transfer of n-alkanes from water to a nonpolar medium [39].

Substance	ΔH^o_{tr} kcal/mol	ΔS^o_{tr} eu	ΔG^o_{tr} kcal/mol
Methane	2.56±0.3	17.6±0.8	−2.86±0.3
Ethane	1.72±0.55	18.1±1.3	−3.59±0.3
Propane	1.77±0.3	22.4±1.1	−4.90
Butane	0.84±0.12	22.3±0.4	−5.94±0.1

4. Although these data provided the starting point for the various iceberg theories of hydrophobic bonding, they agree poorly with values calculated from the theory [39]. In particular, they do not provide a good empirical confirmation for the prediction that changes in ΔG^o_{tr} produced by the addition of a methylene group should arise predominantly from the $T\Delta S^o_{tr}$ term.

To some degree these difficulties have been recognized [35], perhaps even anticipated in that Kauzmann's discussion [38] of hydrophobic bonding includes a warning to the effect that trends in the observed properties of short-chain molecules cannot necessarily be extrapolated to long-chain species. However, this warning like previous efforts [35] to rationalize the data of Table I is based on the idea that the longer alkyl chains may coil up in water, reducing the amount of hydrocarbon-water contact. Several lines of evidence suggest that in fact such coiling becomes important only when the chain contains more than 16 carbon atoms and hence cannot account for the results presented above. The arguments may be summarized as follows:

First, coiling cannot cause a significant reduction in hydrocarbon-water contacts for chains containing less than 5 atoms, but as noted above the average contribution of an added methylene group to ΔH^o_{tr} in the sequence methane, ethane, propane, butane is equal to that found for hexyl to dodecyl derivatives.

Secondly, when partition coefficients K_p between heptane and aqueous buffer were measured for the carboxylic acids $CH_3(CH_2)_{n-2}COOH$, log K_p was found to vary linearly with n for $8 \leq n \leq 16$, but the values for n=16 and n=18 were nearly equal [42]. The dimerization constants of these acids in water also increase smoothly up to, but not beyond n=16. It was suggested [4,42] that this reflects onset of coiling for chains with n>16, which of course requires that coiling must be unimportant for the lower members of the series.

Thirdly, fluorine chemical shifts have been reported for aqueous solutions of many compounds of the type $CF_3(CH_2)_nX$ where $4 \leq n \leq 12$ and X is a solubilizing group which may be non-ionic, anionic, or cationic [5,10-12,43]. At concentrations well below the cmc all these materials produce the same shift, indicating that the average conformation is sufficiently extended to prevent contact between trifluoromethyl groups and the headgroups, and between trifluoromethyl groups and methylene groups some distance along the chain. That coiling does occur for compounds with longer chains, and that this causes an observable change in the position of the fluorine signal, is shown by measurements [44] on $[CF_3(CH_2)_{11}]_2N(CH_3)_2Br$, which may be regarded as a 25-atom chain with the nitrogen atom in the center. So far this is the only known material with a trifluoroalkyl chain which produces an aqueous monomer shift appreciably different from the normal value. The difference of 0.9 ppm at 35° disappears when ethanol replaces water as the solvent, strongly suggesting that indeed the origin of the unusual shift is a conformational change tending to reduce hydrocarbon-water contacts.

Fourthly, water proton chemical shifts for sub-cmc solutions of sodium n-alkyl sulfates show [33] that the effect of an added methylene group is larger when the number of carbon atoms is greater than 6 than when it is less, but in the same direction. By the coiling hypothesis, the change from C_2 to C_4 should create more new hydrocarbon-water contact than the change from C_6 to C_8, and hence produce a larger shift of the proton resonance. A surprising feature of these data, further discussed below, is that in every case increase of the chain length shifts the water peak to higher fields, although the anticipated increase in the average degree of hydrogen bonding for neighboring water molecules would lead one to expect a shift in the opposite direction.

Finally, it has been found that the methylene proton relaxation times in the nmr spectra of sub-cmc solutions of alkyl sulfates in D_2O are very nearly the same as those for solutions of the corresponding alcohols in carbon tetrachloride [34]. These relaxation times are determined by intramolecular proton-proton interactions and depend on the effective rates of internal rotation about the carbon-carbon bonds in the alkyl chain, which must therefore be very similar in both systems. It had been proposed earlier that coiling must be accompanied by a great reduction in the internal motional freedom of a hydrocarbon chain, so that a sizeable difference in the relaxation rates would be anticipated.

Although perpaps none of these arguments would be conclusive by itself, when taken together they seem to provide adequate grounds for the inference that the extent of coiling is not significantly greater in aqueous than in nonaqueous solutions for compounds up to C_{16}. One must then seek to modify prevailing views on the nature of hydrophobic interactions in some other way in order to account for the behavior of alkyl groups in the C_3 to C_{16} range. Perhaps the best starting point for such an endeavor is provided by the models for water-structure effects which stress the analogy between aqueous solutions of small nonpolar molecules and the crystalline clathrate hydrates formed by these materials [45-47].

In the clathrate hydrates, each nonpolar molecule occupies a polyhedral cage of water molecules, with the latter forming a completely hydrogen bonded network. The commonly found polyhedra have 12, 14, or 16 pentagonal or hexagonal faces. The dodecahedron has water molecules at its 20 vertices, the tetrakaidecahedron and hexakaidecahedron have 24 and 28 vertices, respectively, and the free volumes are about 70, 100, and 160 A^3, respectively [48,49]. The observed values of ΔH^o_{tr} and ΔS^o_{tr} for methane, ethane, and propane can be rationalized by assuming that surrounding water molecules become partially ordered through the formation of similar polyhedra, even though in solution any such structures must be very short-lived.

Small polar molecules may act as hydrophobic structure-makers in a very similar way. Mixtures of water with acetone, ethylene oxide, and a number of other organic compounds can form solids exactly analogous to the inert gas hydrates. In dilute solution, these materials cause a downfield shift in the proton nmr signal of the solvent water [50,51], just as one would expect if their effect was to enhance water-water hydrogen bonding. This observation has an important bearing on the interpretation of the shifts caused by dissolved sodium alkyl sulfates and cited above, since it rules out the suggestion [33] that the observed upfield shifts reflect increased hydrogen bonding accompanied by a supposed increase in the covalent nature of the hydrogen bond which reverses the normally expected direction of the chemical shift change.

Ionic compounds like $(n-C_4H_9)_4NF$ also form solid clathrates in which all water molecules are hydrogen bonded, but with the N^+ and F^- atoms replacing oxygen atoms of water in the hydrogen bonded network. The complete hydration shell around each cation consists of four neighboring polyhedra, one for each alkyl chain, with the nitrogen atom at their common vertex. However, chemical shift [52] and X-ray diffraction [53] data show that in solution these clathrate-like polyhedra do not form about alkylammonium ions as effectively as about small uncharged species.

Butane and the larger n-alkanes do not form crystalline hydrates, presumably because the molecules are too large to fit into available polyhedral cavities. Trials with molecular models show that one cannot build water polyhedra with free volumes much over 160 A^3 without either subjecting the hydrogen bonds to severe strain or allowing some broken hydrogen bonds to point toward the interior of the cavity. It is suggested that this is the basic reason why, in solution, the larger alkanes and their water-soluble derivatives are much less effective structure makers than data for the lower homologs would suggest, and may even be structure-breakers.

For example, one might suppose that when an ion like dodecyl sulfate is dissolved in water the hydration near the ionic end is determined primarily by the electrostatic effect of the sulfate group. At the far end of the molecule one may envision transient formation of partial cage structures resembling portions of the polyhedra which enclathrate the lower alkanes. <u>No particularly stable configuration is available for water adjacent to methylene groups in the middle of the chain</u>, and it should have thermodynamic properties not greatly different from those for bulk water. This would account for the fact that increasing the chain length has very little effect on the value of ΔS_m^o or ΔS_{tr}^o for materials with chains of 6 to 16 atoms. The observed increment of -0.6 kcal/mol per methylene group in ΔH_m^o or ΔH_{tr}^o may then serve to indicate that methylene groups in the middle of a chain experience more favorable dispersive van der Waals interactions in a relatively nonpolar medium than in water [39]. The net effect of the dodecyl sulfate ion on the surrounding water could well be a slight reduction in the average degree of hydrogen bonding, in keeping with the most straightforward interpretation of the proton chemical shift changes [33].

If this view of the behavior of water adjacent to hydrocarbon chains is correct, accepted ideas about the changes in thermodynamic variables involved in hydrophobic interactions retain their validity as long as only small alkyl residues are involved, for example in the context of protein structures. However, in the formation of micelles, lipid bilayers or vesicles, and perhaps lipid-protein complexes, one should not expect the hydrocarbon contribution to the free energy change to be dominated by the $T\Delta S^o$ term. It may be that the frequent occurrence of C_{16} to C_{18} chains in lipids reflects the fact that the enthalpic contribution to hydrophobic interactions is maximal for chains of this size, whereas still longer chains would relieve hydrophobic stresses by coiling [42].

ACKNOWLEDGEMENT

This work was supported through Grant GP 19551 by the National Science Foundation.

REFERENCES

1. E. H. Cordes and R. B. Dunlap, Accounts Chem. Res., 2, 329 (1969).

2. E. J. Fendler and J. H. Fendler, Advan. Phys. Org. Chem., 8, 271 (1970).

3. P. Mukerjee and K. J. Mysels, J. Amer. Chem. Soc., 77, 2937 (1955).

4. P. Mukerjee, Advan. Colloid Interface Sci., 1, 241 (1967).

5. N. Muller and R. H. Birkhahn, J. Phys. Chem., 71, 957 (1967).

6. J. A. Pople, W. G. Schneider, and H. J. Bernstein, "High Resolution Nuclear Magnetic Resonance," McGraw-Hill Book Co., New York, N. Y., 1959, p. 222.

7. T. Nakagawa and K. Tori, Kolloid-Z. Z. Polym., 194, 143 (1964).

8. J. C. Eriksson and G. Gillberg, Acta Chem. Scand., 20, 2019 (1966).

9. H. Inoue and T. Nakagawa, J. Phys. Chem., 70, 1108 (1966).

10. N. Muller and T. W. Johnson, J. Phys. Chem., 73, 2042 (1969).

11. N. Muller and F. E. Platko, J. Phys. Chem., 75, 547 (1971).

12. N. Muller, J. H. Pellerin, and W. W. Chen., J. Phys. Chem., to be published. 76 3012 (1972)

13. T. Yasunaga, H. Oguri, and M. Miura, J. Colloid Interface Sci., 23, 352 (1967).

14. E. Graber, J. Lang, and R. Zana, Kolloid-Z.Z. Polym., 238., 470 (1970).

15. E. Graber and R. Zana, Kolloid-Z. Z. Polym., 238, 479 (1970)

16. N. M. Atherton and S. J. Strach, J. C. S. Faraday II, 68, 374 (1972).

17. G. C. Kresheck, E. Hamori, G. Davenport, and H. A. Scheraga, J. Amer. Chem. Soc., 88, 246 (1966).

18. B. C. Bennion, L. K. J. Tong, L. P. Holmes, and E. M. Eyring, J. Phys. Chem., 73, 3288 (1969).

19. B. C. Bennion and E. M. Eyring, J. Colloid Interface Sci., 32, 286 (1970).

20. J. Lang and E. M. Eyring, J. Polymer Sci. A-2, 10, 89 (1972).

21. J. Lang, J. J. Auburn, and E. M. Eyring, J. Colloid Interface Sci., submitted. The author is indebted to Professor Eyring for a prepublication copy of this article.

22. N. Muller, J. Phys. Chem., submitted. 76, 3017 (1972)

23. W. L. Courchene, J. Phys. Chem., 68, 1870 (1964).

24. A. S. Waggoner, A. D. Keith, and O. H. Griffith, J. Phys. Chem., 72, 4129 (1968).

25. O. H. Griffith and A. S. Waggoner, Accounts Chem. Res., 2, 17 (1969).

26. M. Shinitzky, A. C. Dianoux, G. Gitler, and G. Weber, Biochemistry, 10, 2106 (1971).

27. J. E. Gordon, J. C. Robertson, and R. L. Thorne, J. Phys. Chem., 74, 957 (1970).

28. S. J. Rehfeld, J. Phys. Chem., 75, 3905 (1971).

29. J. H. Fendler and L. K. Patterson, J. Phys. Chem., 75, 3907 (1971).

30. E. J. Fendler, C. L. Day, and J. H. Fendler, J. Phys. Chem., 76, 1460 (1972).

31. L. Benjamin, J. Phys. Chem., 70, 3790 (1966).

32. J. M. Corkill, J. F. Goodman, and T. Walker, Trans. Far. Soc., 63, 768 (1967).

33. J. Clifford and B. A. Pethica, Trans. Far. Soc., 60, 1483 (1964).

34. J. Clifford, Trans. Far. Soc., 61, 1276 (1965).

35. L. Benjamin, J. Phys. Chem., 68, 3575 (1964).

36. J. M. Corkill, J. F. Goodman, and J. R. Tate, Trans. Far. Soc., 60, 996 (1964).

37. J. M. Corkill, J. F. Goodman, and J. R. Tate, "Hydrogen Bonded Solvent Systems," A. K. Covington and P. Jones, Eds., Taylor and Francis, Ltd., London, 1968, p. 181.

38. W. Kauzmann, Advan. Protein Chem., 14, 1 (1959).

39. G. Nemethy and H. A. Scheraga, J. Chem. Phys., 36, 3401 (1962).

40. D. C. Poland and H. A. Scheraga, J. Phys. Chem., 69, 2431 (1965).

41. J. M. Corkill and J. F. Goodman, Advan. Colloid Interface Sci., 2, 297 (1969).

42. P. Mukerjee, J. Phys. Chem., 69, 2821 (1965).

43. T. W. Johnson, Ph. D. Thesis, Purdue University, 1970.

44. N. Muller and J. H. Pellerin, Abstracts of the 162nd National Meeting, American Chemical Society, Washington, D. C. Sept. 1971.

45. W. F. Claussen and M. F. Polglase, J. Amer. Chem. Soc., 74, 4817 (1952).

46. D. N. Glew and E. A. Moelwyn-Hughes, Disc. Far. Soc., 15, 150 (1953).

47. D. N. Glew, J. Phys. Chem., 66, 605 (1962).

48. W. C. Child, Jr., Quart. Rev., 18, 321 (1964)

49. G. A. Jeffrey, Accounts Chem. Res., 2, 344 (1969)

50. D. N. Glew, H. D. Mak, and N. S. Rath, Chem. Commun., 264 (1968).

51. D. N. Glew, H. D. Mak, and N. S. Rath, "Hydrogen Bonded Solvent Systems," A. K. Covington and P. Jones, Eds., Taylor and Francis, Ltd., London, 1968 p. 195.

52. H. G. Hertz and W. Spalthoff, Z. Elektrochem. 63, 1096 (1959).

53. A. H. Narten and S. Lindenbaum, J. Chem. Phys., 51, 1108 (1969).

MICELLAR CATALYSIS FOR CARBONIUM ION REACTIONS

Jan Baumrucker and Maritza Calzadilla

Escuela de Quimica, Universidad Central

Caracas, Venezuela

E. H. Cordes

Department of Chemistry, Indiana University

Bloomington, Indiana 47401

INTRODUCTION

During the past 15 years, there has been an increasing level of activity in the study of the catalysis of organic reactions by micelles [1-3]. In the course of these studies, a number of types of reactions have come under scrutiny, several of which are discussed in some detail in later pages of this volume. Our own efforts have centered largely on certain reactions which involve carbonium ions either as substrates or reaction intermediates.

A very large number of carbonium ion reactions are known in organic chemistry; these species are frequently encountered as intermediates in solvolyses, rearrangements, and decarboxylations, among other reactions. Moreover, carbonium ions are likely to be involved in many enzyme-catalyzed reactions. Although definite evidence for the formation of carbonium ions in the course of nonenzymatic reactions can frequently be obtained, it is generally necessary to rely on more speculative arguments for enzymatic cases. However, in the case of lysozyme-catalyzed hydrolysis of polysaccharides, eq. 1, secondary deuterium isotope

$$\text{(cyclohexane-OR)} \xrightarrow{\text{slow}} \text{[O=C}_1\text{]}^+ + {}^{\ominus}\text{OR} \xrightarrow[\text{fast}]{H_2O} \text{(cyclohexane-OH)} \tag{1}$$

effects provide strong evidence for carbonium ion formation in the transition state [4]. Consequently, understanding of mechanism and catalysis for carbonium ion reactions will prove significant for both organic and biochemistry.

Let us begin our discussion of micellar catalysis for carbonium ion reactions by asking what we might expect to observe, particularly in the case of micelles formed from ionic surfactants. As is developed below, these reactions will occur at or near the micellar Stern layer, some of the properties of which have been discussed by Muller in the previous chapter. Briefly, the surface of an ionic micelle possesses a high density of charged groups, some of which are neutralized by incorporation of counterions into the Stern layer, in a region of relatively low polarity. The most reliable estimates suggest that the micellar surface has a polarity near that for pure ethanol [3,5]. It may be a unique character of the micellar surface that such a high density of charged groups are found in an environment of this polarity. As developed by Muller in the previous chapter, solvent water molecules very probably penetrate beyond the polar headgroups at the micellar exterior. Consequently, solutes in the water phase may interact both with the nonpolar chains of the individual surfactant molecules and with the polar headgroups, without necessarily penetrating into the nonpolar micellar core. Thus, the micellar surface may be referred to as amphipathic, having affinity for both polar and nonpolar species [3]. In fact, as we will see later in this chapter and in later ones, many reactions catalyzed by ionic micelles involve one nonpolar reactant and one ionic one.

There is a very large electrical potential drop across the micellar surface, amounting to several tens of thousands of volts per centimeter. This reflects both the high concentration of unneutralized charges at the micellar surface and the small thickness of the surface region itself so that the potential drop occurs over a very short distance. As an initial approximation, we might expect to be able to predict the qualitative behavior of reaction kinetics in ionic micellar systems on the basis of differences in electrostatic stabilization of ground state, transition state, and product state by the micellar surface. Let us follow this line or argument for three types of reactions involving carbonium ions:

$$(R-X)_m \rightarrow R_m^+ + X_m^- \qquad (2)$$

$$(R-X)_m + H_b^+ \rightarrow R_m^+ + HX_m \qquad (3)$$

$$R_m^+ + X_b^- \rightarrow (R-X)_m \qquad (4)$$

In these three equations, the subscripts m and b have been employed to denote the micellar and bulk phases, respectively.

In eq. 2, a simple unimolecular decomposition reaction is depicted for a substrate localized on a micelle; the immediate products, a carbonium ion and an anion, will also be produced at the micellar surface, although one or both may subsequently diffuse into the bulk phase prior to undergoing further reactions. In this simple case, electrostatic effects at the micellar surface will be opposing for the two immediate products: an anionic surface expected to stabilize the carbonium ion would be expected to destabilize the anionic product and vice versa. Consequently, it is difficult to predict if one should expect catalysis or inhibition for such reactions by ionic surfactants. In fact, no studies of such reactions in micellar systems have been carried out.

The reaction in eq. 3 differs from the preceding one in that acid-catalysis is involved. The crucial distinction is that the leaving group now departs as a neutral species rather than as an anion. The electrostatic situation at the micellar surface is now much simpler: an anionic surface would be expected to stabilize the carbonium ion relative to the uncharged reactant and a cationic surface should have the opposite effect. Note that this argument depends on the ground state for the catalyzing acid being the bulk phase, as it ordinarily is. Consequently, we expect that anionic surfactants should catalyze and cationic surfactants inhibit reactions in this class. There is a great deal of evidence to support this view and we devote the bulk of the remainder of this chapter to a detailed analysis of it.

In eq. 4, the attack of an anion from the bulk phase on a carbonium ion localized in the micelle is viewed. Note carefully that this reaction is not the reverse of that shown in eq. 2. The crucial difference is that the reverse of the latter process involves attack of an anion associated with the micelle, not one free in the aqueous milieu. In the case of the reaction type of eq. 4, the qualitative arguments based on electrostatic considerations are straightforward. For a cationic micelle, the cationic substrate should be electrostatically destabilized relative to the uncharged transition state. Consequently, catalysis of such reactions by cationic surfactants is anticipated and, as developed below, is found. Note finally that if the anionic nucleophilic reagent in eq. 4 is replaced by a neutral one, both substrate and transition state possess a full positive charge and catalysis by cationic surfactants is not predicted.

Electrostatic considerations of the type just developed are by no means the only factors which influence reaction rates in micellar systems. In some cases they may not even be the most important ones. But in most cases, the qualitative effects on reaction rate, inhibition, catalysis, or absence of a change, can be accounted for on this basis. With this background information in hand, let us move now to consider experimental results on two systems: first, hydrolysis of acetals, ketals, and ortho esters (eq. 3) and, second, attack of nucleophiles on stable carbonium ions (eq. 4).

MICELLAR CATALYSIS FOR HYDROLYSIS OF ACETALS, KETALS, AND ORTHO ESTERS

The hydrolysis of these substrates has been studied actively for several decades. Consequently, a great deal of information is available concerning mechanisms for these reactions. Prior to examining the micellar catalysis, it is appropriate to pause just long enough to develop the central aspects of the reaction mechanisms for media not containing surface-active materials.

Basic Aspects of the Hydrolysis of Acetals and Related Substrates

Normally, acetals, ketals, and ortho esters hydrolyze at measurable rates in slightly acidic aqueous media via a sequence of reactions involving acid catalysis and the formation of a carbonium ion in the transition state. The overall reaction pathway is:

$$\underset{OR}{\overset{OR}{\diagup\!\!\!\diagdown}}C \xrightleftharpoons{+H^+} \underset{\overset{+OR}{H}}{\overset{OR}{\diagup\!\!\!\diagdown}}C \xrightleftharpoons{-ROH} \overset{OR}{\diagup}C^+ \xrightleftharpoons{+H_2O}$$

$$\underset{OR}{\overset{OH}{\diagup\!\!\!\diagdown}}C \rightarrow \rightarrow \diagup\!\!\!\diagdown C=O + ROH \quad (5)$$

There is definitive evidence to suggest that, in most cases at least, the formation of the initial carbonium ion is irreversible. That is, either substrate protonation, or formation of the carbonium ion (which may be concerted with substrate protonation), or both are involved in the transition state [6]. The various possibilities can be understood in terms of the following equations:

$$S + H_3O^+ \underset{k_{-1}}{\overset{k_1}{\rightleftharpoons}} SH^+ + H_2O$$

$$SH^+ \overset{k_2}{\longrightarrow} products \quad (6)$$

Provided that k_{-1} (H_2O) is much greater than k_2, protonation of the substrate is rapid and reversible and only carbonium ion formation is involved in the transition state. The situation obtains for the hydrolysis of simple acetals and ketals. Should the magnitudes of k_{-1} and k_2 be similar then either (i) the two steps both contribute to the overall rate of the reaction or (ii) the two processes are concerted. In the latter case, simultaneous proton transfer and carbon-oxygen bond cleavage occur in the transition state:

$$\underset{\text{OR}}{\overset{\text{OR}}{\diagdown}}\!\!\!\!\!\!\!\underset{}{\overset{}{\diagup}}\text{C} + H_3O^+ \rightarrow \left[\begin{array}{c} \diagdown \\ -\text{C}---\text{O}---\text{H}---\text{OH}_2 \\ \diagup \ \ \ \ \ | \\ \text{OR} \ \ \ \ \text{R} \end{array} \right]^+ \quad (7)$$

$$\rightarrow \diagdown\text{C}^+\!-\text{OR} + \text{ROH} + H_2O$$

The bulk of the evidence, including observation of general acid catalysis for certain of these reactions, suggests that the latter possibility (eq. 7) more accurately represents the truth. This case ordinarily obtains for the hydrolysis of ortho esters and orthocarbonates [6,7]. It has also been observed for hydrolysis of acetals and ketals which possess either particularly facile leaving groups or form particularly stable carbonium ion intermediates [8,9]. For our subsequent considerations, the important point is that the transition state involves cleavage of a carbon-oxygen bond, which may be preceded by or accompanied by proton transfer, leading to carbonium ion formation. It is the stabilization of this developing carbonium ion in the transition state that accounts for the catalytic effects observed when these reactions are conducted in the presence of anionic surfactants, a matter to which we now turn.

Interaction of Acetals and Related Substrates with Micelles

Prior to any alteration of the rate of hydrolysis of an acetal or related substrate, it is required that this substrate be incorporated into or onto the micelles. An alteration in reaction velocity elicited by addition of ionic surfactants is itself strong evidence that such interaction has occurred but leaves two questions unanswered: what is the equilibrium constant for complexation of the substrate with the micelle? and what is the site of localization of the adsorbed molecule with respect to the micelle? The former question is most important in connection with explanations of the quantitative nature of rate-concentration profiles, a matter considered below. A method developed by Richards and collaborators involving elution of the substrate from columns of molecular seives as a function of the concentration of surfactant provides a means for attacking the first question [10]. For a specific case of interest here, the equilibrium constant for interaction of methyl orthobenzoate with micelles of sodium dodecyl sulfate at 25°

has a value of 73 $\underline{M}^{-1}$, based on the total concentration of surfactant in the system (not on the concentration of micelles) [11]:

$$S + M \rightleftarrows S \cdot M \; ; \; K_{eq} = \frac{(S \cdot M)}{(S)(M)} = 73 \; \underline{M}^{-1} \qquad (8)$$

Thus, at a concentration of sodium dodecyl sulfate (SDS) near 0.014 $\underline{M}$, we calculate that about half of the methyl orthobenzoate ought to be associated with the micelles formed from this surfactant and the other half ought to be present in the bulk solution. The relationship of this value to the kinetics of hydrolysis of this ortho ester in the presence of SDS will become apparent shortly.

Determination of the site of localization of a substrate in a micelle is not trivial. Muller has just described a number of possible approaches to this and related matters including fluorine-19 chemical shift data. This approach has been employed with fluorine-labeled acetals and ortho esters: results are collected in Table 1 [12].

TABLE 1

^{19}F Chemical shifts for methyl ortho-p-fluorobenzoate and p-fluorobenzaldehyde diethyl acetal for several solvent systems (from ref. 12).

Solvent	Chemical Shifts, ppm[a]	
	Methyl-ortho-p-fluorobenzoate	p-Fluorobenzaldehyde diethyl acetal
Hexane	36.42 (25)	37.55 (31)
Decane	37.67	...
Benzene	35.74 (25)	36.68
Pyridine	35.92	36.97
Triethylamine	36.06 (25)	37.34
n-Butylamine	36.49	37.64
N-Methylacetamide	36.22	37.45
2,4-Dimethylphenol	33.95 (26)	35.32
o-t-Butylphenol	34.04 (25)	34.68 (25)
Methanol	36.58	38.33
80% Aq. methanol[b]	36.19 (39)	38.12 (25)
60% Aq. methanol[b]	35.74 (40)	37.48 (25)

Solvent	Chemical Shifts, ppm[a]	
	Methyl-ortho-p-fluorobenzoate	p-Fluorobenzaldehyde diethyl acetal
53% Aq. methanol[b]	35.64 (100)	...
47% Aq. methanol[b]	35.53 (100)	...
40% Aq. methanol[b]	...	36.72 (100)
30% Aq. methanol[b]	35.74 (768)	36.38 (435)
15% Aq. methanol[b]	35.23 (1347)	...
Dimethylformamide	...	37.82 (16)
Dimethyl sulfoxide	...	36.13
Acetone	...	38.44
0.1 M SDS[b,c]	33.88 (150)	35.82 (87)
0.1 M SDS, 0.01 M substrate	34.02 (110)	35.43 (1125)
0.075 M SDS, 0.0075 M substrate	34.10 (223)	35.57 (2106)
0.05 M SDS, 0.005 M substrate	34.02 (200)	...
0.10 M SDS, 0.005 M substrate	33.87 (301)	...

[a]All chemical shifts are upfield with respect to an external standard of neat trifluoroacetic acid. Number of sweeps of each spectrum is given in parentheses following chemical shift for those cases in which more than one sweep was employed. See the text for details of the measurement process. [b]Measured in the presence of 0.042 M sodium carbonate buffer. [c]SDS is sodium dodecyl sulfate. Measured in the presence of saturating concentrations of the substrates.

Detailed analysis of this data is difficult, but it appears that there is an important contribution to the chemical shift from hydrogen bond formation from protic solvent molecules to the organic substrates. Comparison of the chemical shifts for the organic solvent data and the data obtained in the presence of SDS suggests that substrates incorporated into the micelles still form hydrogen bonds with water; that is, that they are not hidden from the bulk solvent through incorporation into the inner core of the micelle. This conclusion is supported by measurement of shielding parameters [12].

Electrostatic Factors in Micellar Catalysis

Arguments developed earlier suggest that micelles formed from anionic surfactants should catalyze the hydrolysis of acetals and related substrates through electrostatic stabilization of the developing carbonium ion in the transition state. By the same reasoning, we expect that micelles formed from cationic surfactants will inhibit hydrolysis of these substrates. There is abundant evidence to support these expectations. In Table 2, are collected several examples of the qualitative effect of a number of surfactants on the rate of hydrolysis of several acetals, ketals, and ortho esters. Note that in every case studied, anionic surfactants are catalytic and cationic ones inhibitory, despite the substantial variation in both surfactant and substrate structure. Many factors of the catalyzed reactions depend in a sensitive way on details of both surfactant and substrate structure, as is developed below, but the gross qualitative behavior does not.

Examination of the data in Table 2 reveals that non-ionic and zwitterionic surfactants inhibit hydrolysis of these substrates. This result is probably the consequence of the lower polarity at the micellar surface compared to the bulk phase; it has been established that ortho ester hydrolysis is inhibited by addition of solvents of low dielectric constant [6].

Surfactant Concentration-Rate Profiles

One of the fundamental characteristics of micelle-catalysis for organic reactions is the nature of the dependence of reaction rate on surfactant concentration. A typical example, for the SDS-catalyzed hydrolysis of methyl orthobenzoate, is shown in Figure 1 [11]. The concentration profile is multiphase; below the cmc for the surfactant, the rate constants are independent of surfactant concentration. Above the cmc, the rate constants rise rapidly with increasing surfactant concentration, level off, and finally decrease with increasing concentration of this surfactant. At the optimal surfactant concentration, a rate augmentation of 85-fold at 25° is observed for this reaction. Profiles of this type can be rationalized on the basis of (1) the necessity of micelles for catalysis, (2) adsorption of a progressively greater fraction of the substrate into the

TABLE 2

A summary of qualitative effects of several surfactants on the rate and hydrolysis of acetals, ketals, and ortho esters.

Substrate	Surfactants	Effect	Reference
Methyl Orthobenzoate, $C_6H_5C(OCH_3)_3$	Sodium Alkylsulfates	⎫	11,13,14
	Isomeric Hexadecylsulfates	⎪	
	Substituted Oxyethylenesulfates	⎬ Catalysis	
	Sodium 2-Dodecylbenzenesulfonate	⎪	
	Disodium Sulfoalkylsulfates	⎭	
	Alkyldisulfonates		
	Sulfoalkylcarboxylates		
	α-Sulfoalkyl Esters		
	Dodecyldimethylphosphine Oxide	⎫	
	Dodecyldimethylammoniumpropane-sulfonate	⎬ Inhibition	
	Dodecyldimethylammoniumacetate	⎪	
	Cetyltrimethylammoniumbromide	⎭	
p-Substituted Methyl Orthobenzoates, $X\text{-}C_6H_4\text{-}C(OCH_3)_3$ $X = -OMe, -Me, -H, -F, -Cl, -NO_2$	Sodium Dodecylsulfate	Catalysis	13

Substrate	Surfactants	Effect	Reference
p-Substituted Benzaldehyde Diethyl Acetals X-C6H4-CH(OEt)2 X = -OMe, -Me, -H, -F, -Cl	Sodium Dodecylsulfate Sodium Decylsulfonate	Catalysis	15,16
2-(p-Substitutedphenoxy)-tetrahydropyrans X = -OMe, -Me, -H, -Cl, -NO2	Sodium Dodecylsulfate Sodium 2-Hexadecyloxyethylsulfate Sodium 2-Hexadecyloxy-1-methylsulfate	Catalysis	17
	Dodecyldimethylammoniumpropanesulfonate Dodecyldimethylphosphine Oxide	Inhibition	
Benzophenone Dimethyl Ketal (C6H5)2C(OCH3)(OOH3)	Sodium Dodecylsulfate Octadecyldimethylammonium Bromide	Catalysis Inhibition	18

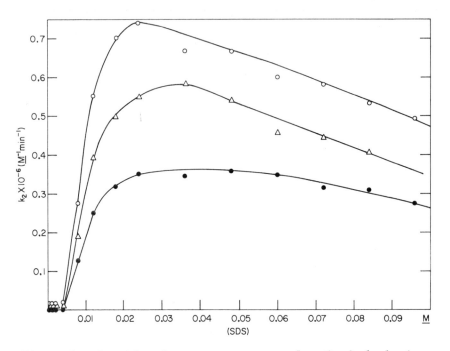

Figure 1. Second-order rate constants for the hydrolysis of methyl orthobenzoates in aqueous solution at 40° (o), 32.5° (△), and 25° (●), plotted against the concentration of sodium dodecyl sulfate. Values of pH were maintained through the addition of 0.01 M acetate buffers.

micellar phase until that fraction approaches unity with increasing surfactant concentration, and (3) inhibition of the micellar reaction by the counterions of the surfactant itself. In the case of methyl orthobenzoate hydrolysis in the presence of sodium dodecyl sulfate, it has been shown that this interpretation must be substantially correct [11]. As noted above the equilibrium constant for the association of substrate with surfactant is 73 M^{-1}. This value accounts quantitatively for the increase in rate constant with increasing surfactant concentration. That is, when the substrate is predicted to be 50 percent associated with the micellar phase on the basis of the equilibrium constant, about 50 percent of the maximum catalysis is experienced and so on. Furthermore, when the total concentration of sodium ion is maintained constant by the addition of the necessary quantities of inorganic salts, the inhibition of the reaction at high surfactant concentrations disappears.

By neglecting the salt inhibition at high surfactant concentrations, it is possible to construct an approximate model which accounts semiquantitatively for the shape of the concentration-rate profile. Generally, this model will have the form:

$$M + S \xrightleftharpoons{K} M \cdot S$$
$$\downarrow k_w \qquad \downarrow k_m \qquad (9)$$
$$\text{products} \quad \text{products}$$

in which $\underline{M}$ is the micelle, $\underline{S}$ is the substrate, k_w and k_m are the rate constants for the reaction in the bulk and the micellar phase, respectively, and $\underline{K}$ is the equilibrium constant for association of the substrate with the micelle. Bunton and his co-workers have provided a good example of the use of this model in studies on the kinetics of hydrolysis of 2,4-, and 2,6-dinitrophenyl phosphates [19].

Substrate Concentration-Rate Profiles

By analogy with other systems, including enzymatic ones, in which a complex is formed between two reactants prior to bond-changing reactions, one might expect that saturation of the micellar phase with increasing substrate concentrations would be observed. In certain cases at least, such behavior is found. In Figure 2, the first-order rate constants for hydrolysis of methyl orthobenzoate in the presence of 0.001 $\underline{M}$ sodium dodecyl sulfate are plotted as a function of the concentration of the ortho ester [14]. The decreasing rate with increasing substrate concentration most likely represents saturation of the micellar phase with substrate. Thus as substrate concentration increases beyond the saturation point, an increasing fraction of the substrate must exist free in the solution. As this fraction approaches unity, the rate constant for the reaction must approach that for the reaction in purely aqueous solutions, as indeed it does.

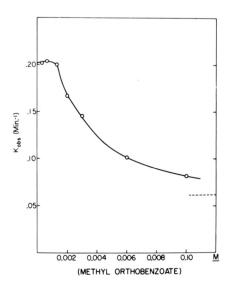

Figure 2. First-order rate constants for the hydrolysis of methyl orthobenzoate in the presence of 0.001 M sodium dodecyl sulfate at 25° and pH 4.95 plotted as a function of the substrate concentration. The dotted line indicates the rate constant under these conditions in the absence of surfactants.

The Effect of Surfactant Structure

Quite aside from changes in surfactant structure which involve alterations in charge type, there are two general areas of surfactant variability: the nature and position of the head group and the size of the hydrophobic chain. Both factors are important for the micellar catalysis of acetals and ortho esters.

In Figure 3, are plotted rate constants for hydrolysis of methyl orthobenzoate as a function of the concentration of several sodium hexadecyl sulfates [13]. Note that catalytic efficiency is markedly dependent on the position of the head group: as it is moved further from the terminus of the chain, the catalytic efficiency decreases, both in terms of the maximal catalysis achieved and in terms of the concentration of surfactant necessary to achieve it.

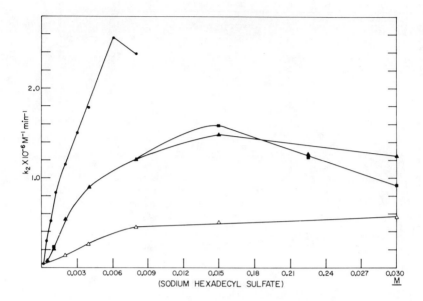

Figure 3. Second-order rate constants for hydrolysis of methyl orthoebnzoate plotted against the concentration of 1-hexadecyl (●), 2-hexadecyl- (■), 3-hexadecyl- (▲), and 5-hexadecyl (△) sodium sulfate.

In addition to the position of the head group, the nature of the charge bearing group influences catalytic effectiveness for the hydrolysis of methyl orthobenzoate, even though the charge type itself does not change. An example is provided in Figure 4. Detailed analysis of this data would require a great many more experiments but the sensitive dependence of the degree of catalysis on the details of surfactant structure is clear. Note that those surfactants bearing two negative charges are particularly effective catalysts. Only 0.0064 $\underline{M}$ disodium 2-sulfo-2-methyloctadecanoate causes more than a forty-fold increase in the rate of hydrolysis of the ortho ester [13].

Finally, let us focus attention on the role of surfactant hydrophobicity in determination of catalytic efficiency. It is quite generally true that the more hydrophobic the surfactant, the better catalyst it is. In Table 3, some rate increases for methyl orthobenzoate hydrolysis elicited by several sodium alkyl sulfates are collected. Note that both the maximal catalysis observed and the surfactant

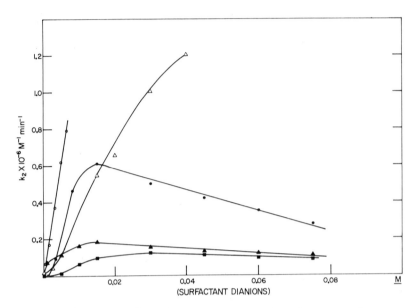

Figure 4. Second-order rate constants for hydrolysis of methyl orthobenzoate plotted against the concentration of disodium 2-sulfo-2-methyloctadecanoate (o), disodium-2-sulfo-2-butyltetradecanoate (Δ), disodium 2-sulfooctadecyl sulfate (●), disodium 2-sulfoethyl α-sulfostearate (▲), and disodium 2-sulfoethyl α-sulfopalmitate (■).

concentration required to elicit maximal catalysis are sensitive functions of the length of the alkyl chain [11]. This behavior may reflect the partitioning of substrate between micellar bulk phase or the precise localization of substrate with respect to the micellar surface which may lead to a direct contribution of hydrophobic forces to the activation energy for the reactions. It is not possible to assess the relative contribution of each of these factors to the observed rate effects at this time.

Effect of Substrate Structure on Micellar Catalysis

There are two aspects of substrate structure that have proved interesting in terms of reaction kinetics in micellar systems: substrate hydrophobicity and polar substituents. Generally, increasing the hydrophobic character of the substrate increases the influence of the micellar phase on the

TABLE 3

Maximal rate increases elicited by a series of sodium alkyl sulfates for methyl orthobenzoate hydrolysis.

Sodium alkyl sulfate	Temp. °C	$k_2^0 \times 10^{-6}$ [a] M^{-1} min^{-1}	$k_2 \times 10^{-6}$ [b] M^{-1} min^{-1}	Max rate increase
Octyl	25.0	0.00502	0.0351 at 0.20 $\underline{M}$	7.0
Decyl	25.0	0.00452	0.121 at 0.075 $\underline{M}$	26.8
Dodecyl	25.0	0.00452	0.357 at 0.048 $\underline{M}$	79.0
Tetradecyl	30.0	0.00864	0.793 at 0.20 $\underline{M}$	91.8

[a] Second-order rate constants in the absence of surfactant. [b] Second-order rate constants for the reaction in the presence of the indicated concentrations of surfactants at which values maximum catalysis occurs.

velocity of the reaction, just as increasing the hydrophobicity of the surfactant tends to accentuate these effects. This is true for hydrolysis of acetals and ortho esters. For example, hydrolysis of methyl orthobenzoate and methyl orthovalerate are subject to catalysis by SDS but hydrolysis of methyl orthoacetate is not [14]. These results most likely reflect the partitioning of substrate between the micellar and bulk phases. For other reactions, however, it seems likely that hydrophobic interactions may contribute to activation energies [20].

The effects of polar substituents on the rates of organic reactions in micellar phases appear to differ substantially from such effects in aqueous solution. Some pertinent data for hydrolysis of acetals and ortho esters is collected in Table 4. Note that, in each case, the reaction in the micellar phase is more sensitive to the nature of polar substituents than the same reaction in water. This data may reflect either medium effects or changes in transition state structure. No firm basis exists for distinguishing between these alternatives at this time.

In still other cases, the effect of polar substituents on reaction rates in micellar systems is more complicated. Thus, rate constants for hydrolysis of a series of substituted phenoxytetrahydropyrans in the presence of sodium dodecyl sulfate [17] fail to generate a satisfactory Hammett

TABLE 4

Effects of polar substituents on the hydrolysis of acetals and ortho esters in water and in the presence of anionic surfactants.

Substrates	Surfactant	$\rho_{micellar}$	$\rho_{aqueous}$	Ref.
Methyl orthobenzoates	sodium dodecyl sulfate	-2.5	-1.2	13,21
Benzaldehyde diethyl acetals	sodium dodecyl sulfate	-4.1	-3.3	15
Benzaldehyde diethyl acetals	sodium dodecyl sulfonate	-4.3	-3.2	22

plot although the same reactions in aqueous solution do [23,17]. This results presumably reflects an effect of polar substituents on the precise localization of the substrates with respect to the micellar surface.

$$\text{(structure)} \xrightarrow{H_2O} \text{(products)} \quad (10)$$

Salt Effects

One of the striking aspects of the kinetics of organic reactions in micellar systems is their sensitivity to salt effects. Changes in the nature or concentration of electrolyte that would lead to barely detectable differences in rates of reactions in purely aqueous systems frequently cause differences of an order of magnitude or more for the same reactions in the presence of ionic surfactants. As an example, we have already noted the inhibition of surfactant-dependent reactions due to the counterion of the surfactant itself.

More systematic studies have been performed: the results of two such studies are shown in Figures 5 and 6. Here second-order rate constants for hydrolysis of methyl orthobenzoate in the presence of sodium dodecyl sulfate are plotted as a function of several cations. All are markedly

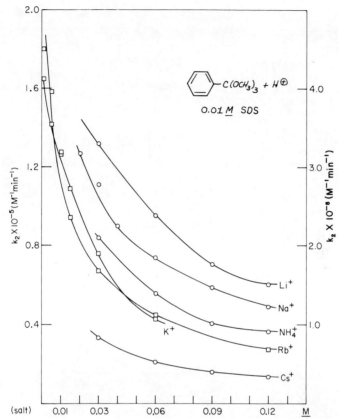

Figure 5. Second-order rate constants for the hydrolysis of methyl orthobenzoate in the presence of 0.01 M sodium dodecyl sulfate plotted against the concentrations of several alkali metal ions and ammonium ion.

inhibitory. Note that for both series of ions the inhibition increases as the hydrophobic character of the salt increases. These observations can be readily understood in terms of increasing the extent of charge neutralization of the micellar surface. To the extent that catalysis is dependent on electrostatic stabilization of the transition state with respect to the ground state, such charge neutralization must reduce the catalytic effect. In other cases, the salt inhibition may derive principally from the displacement of one reactant from the micellar surface by the electrolyte.

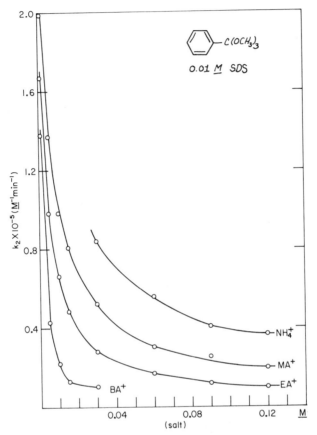

Figure 6. Second-order rate constants for hydrolysis of methyl orthobenzoate plotted as a function of the concentration of several ammonium ions; MA, methyl-; EA, ethyl; and BA, butylammonium ion.

Effects of Organic Additives

Relatively little work has been accomplished concerning the effects of organic additives on the kinetics of surfactant-dependent reactions. Those conclusions that are available indicate that the effects are likely to be complex and to differ from one system to another. For example, in Figure 7, second-order rate constants for the hydrolysis of methyl orthobenzoate in the presence of 0.01 $\underline{M}$ sodium dodecyl

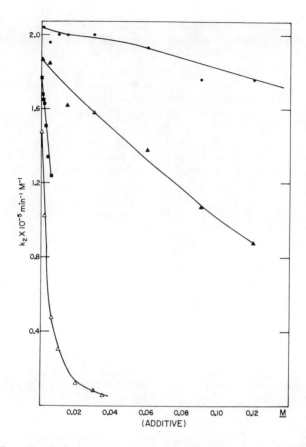

Figure 7. Second-order rate constants for the hydrolysis of methyl orthobenzoate in the presence of 0.01 M sodium dodecyl sulfate at 25° plotted against the concentration of ethanol (●), n-butanol (▲), n-heptanol (■), and dimethyldodecylphosphine oxide (△).

sulfate solution at 25° are collected as a function of the concentration of ethanol, n-butanol, n-heptanol, and dimethyldodecylphosphine oxide. As is apparent from the figure, the extent of inhibition of the surfactant-catalyzed reaction becomes greater as the concentration of alcohol is increased. Further, the effectiveness of the alcohol at any given concentration as an inhibitor increases with increasing length of the carbon chain of the alcohol molecule. As an example, n-decanol (not shown in the figure) at a concentration of 7.9×10^{-5} M inhibits the reaction about two-fold.

The inhibition of the sodium dodecyl sulfate-dependent hydrolysis of methyl orthobenzoate by these nonionic additives may be the consequence of one or all of the following factors (1) increase in the cmc of the anionic surfactant (particularly in the case of ethanol and n-butanol inhibition), thus lowering the concentration of micelles in the solution and hence the fraction of the ortho ester in the micellar phase; (2) displacement of methyl orthobenzoate from the micellar phase by the additives; or (3) lessening of the electrostatic stabilization of the transition state. There exists no basis for distinguishing between these factors or for assigning relative weights to them at this time.

The foregoing analysis of the various factors that influence rates and equilibria of organic reactions in micellar systems is based on consideration of electrostatic and hydrophobic forces. In some cases that have come to light, the effects observed are not easily assigned to these interactions and may result principally from local medium effects, or perhaps from electrostatic effects of a subtler sort.

MICELLAR CATALYSIS FOR FADING OF CRYSTAL VIOLET

Crystal violet is a stable carbonium ion which is transformed into the corresponding alcohol in aqueous solution in a reaction catalyzed by hydroxide ion:

$$[(CH_3)_2N-C_6H_4]_3C^+ + OH^- \longrightarrow [(CH_3)_2N-C_6H_4]_3C-OH \tag{11}$$

In the presence of surfactants, this reaction corresponds to the formalism expressed in eq. 4, inasmuch as it involves the attack of an anion from the bulk phase on a carbonium ion which is sufficiently hydrophobic to be incorporated into micelles. Our previous discussion would suggest that this reaction be subject to catalysis by cationic surfactants and to inhibition by anionic ones.

These expectations have been fully born out. In fact, in 1959 Duynstee and Grunwald, in the first thorough investigation of reaction kinetics in micellar systems [24], studied this and related reactions. A qualitative summary of their results is shown in Table 5. In each case studied, the behvaior expected on the basis of simple electrostatic arguments is obtained.

In an effort to further our understanding of surfactant catalysis for the addition of hydroxide ion to crystal violet, we have explored some facets of this reaction not developed in the original work. Results obtained are fully consistent with those already presented for the hydrolysis of acetals, ketals, and ortho esters.

In Figure 8 are collected rate-concentration profiles for the alkaline fading of crystal violet in the presence of several n-alkyltrimethylammonium bromides [25]. Note the dramatic effect of the surfactant chain length on the observed kinetics: catalysis by the C_{10} and C_{12} surfactants is observed only at higher surfactant concentrations than those shown in this figure. As the chain is further lengthened, the catalytic efficiency of these surfactants is greatly accentuated. This is perhaps the most dramatic example of the role of hydrophobic interactions in the kinetics of micelle-catalyzed reactions yet observed.

TABLE 5

Effects of micelle-forming surfactants on the kinetics of alkaline fading of several stable carbonium ions.[a]

Dye	pH	No Surf.	10^4 k, sec^{-1} 0.01 M CTAB[b]	0.01 M SDS
Crystal violet	12.00	17	241	1
Malachite green	10.2	8.1	70	1.8
Brilliant green	10.2	5.1	93	---

[a]From the data of E. F. J. Duynstee and E. Grunwald, J. Amer. Chem. Soc., 81, 4540 (1959). [b]Hexadecyltrimethylammonium bromide.

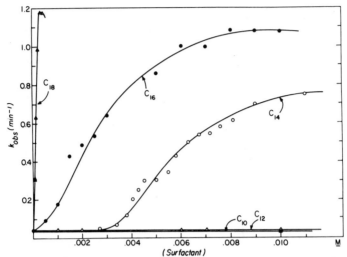

Figure 8. First-order rate constants for the attack of hydroxide ion on crystal violet in aqueous solution at 30°, (OH⁻) = 0.003 M, plotted as a function of the concentration of several n-alkyltrimethylammonium bromides.

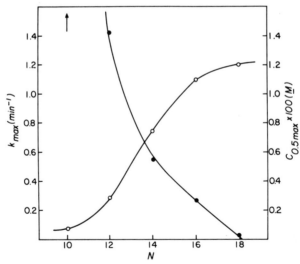

Figure 9. Plots of the maximal first-order rate constants (left scale, open circles) and the surfactant concentrations required to reach half-maximal catalysis (right scale, closed circles) for the fading of crystal violet as a function of the chain length of n-alkyltrimethylammonium bromides.

As is shown in Figure 9, increasing surfactant hydrophobicity affects not only the concentration of surfactant required to achieve maximal catalysis but the extent of total catalysis is achieved as well. It follows that the effect of chain length involves more than variation in the cmc of the surfactant and changes in the equilibrium constant for adsorption of the substrate onto the micelle. The fact that maximal velocities are affected by chain length indicates that, through one mechanism or another, hydrophobic interactions help to determine the free energy of activation for organic reactions.

Finally, we note that salts inhibit the surfactant-catalyzed alkaline fading of crystal violet just as they do in the case of ortho ester hydrolysis. The pertinent data are shown in Figure 10. Qualitatively, the differential effects of the various anions are in accord with expectations based on the affinity of the anions for the micellar surface and with the results observed for inhibition of p-nitrophenyl hexanoate hydrolysis [26]. More specifically, those anions with the highest affinity for the micelles and, hence, possessing the greatest capacity to neutralize the micellar charge, are the best inhibitors. This is identical to the behavior observed in the converse case of cation inhibition for methyl orthobenzoate hydrolysis, Figures 5 and 6.

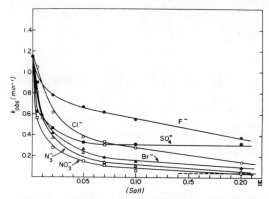

Figure 10. First order rate constants for the fading of crystal violet in aqueous solution containing 0.003 $\underline{M}$ sodium hydroxide and 0.01 $\underline{M}$ hexadecyltrimethylammonium bromide at 30° plotted as a function of the concentration of several anions. The dotted line in the lower right of the figure is the rate constant under these conditions in the absence of surfactant.

Acknowledgement: We are indebted to the American Chemical Society for permission to reproduce Figures 2, 4, 5, 6, 8, 9 and 10 and to the National Institutes of Health for generous support of the work described herein (Grant AM 08232).

REFERENCES

1. E. H. Cordes and R. B. Dunlap, Accts. Chem. Res., 2, 329 (1969).

2. E. J. Fendler and J. H. Fendler, Advan. Phys. Org. Chem., 8, 271 (1970).

3. E. H. Cordes and C. Gitler, Bioorganic Chemistry, 2, 1 (1973).

4. P. W. Dahlquist, T. Rand-Meir, and M. A. Raftery, Proc. Natl. Acad. Sci. U.S., 61, 1194 (1968).

5. P. Mukerjee and A. Ray, J. Phys. Chem., 70, 2144 (1966).

6. E. H. Cordes, Progr. Phys. Org. Chem., 4, 1 (1967).

7. C. A. Bunton and R. A. DeWolfe, J. Org. Chem., 30, 1371 (1965).

8. E. Anderson and T. H. Fife, J. Amer. Chem. Soc., 91, 7163 (1969).

9. T. Fife and L. Brod, J. Amer. Chem. Soc., 92, 1681 (1970).

10. D. G. Herries, W. Bishop, and F. M. Richards, J. Phys. Chem., 68, 1842 (1964).

11. R. B. Dunlap and E. H. Cordes, J. Amer. Chem. Soc., 90, 4395 (1968).

12. P. A. Arrington, A. Clouse, D. Doddrell, R. B. Dunlap, and E. H. Cordes, J. Phys. Chem., 74, 665 (1970).

13. R. B. Dunlap and E. H. Cordes, J. Phys. Chem., 73, 361 (1969).

14. M. T. A. Behme, J. Fullington, R. Noel, and E. H. Cordes, J. Amer. Chem. Soc., 87, 266 (1965).

15. R. B. Dunlap, G. A. Ghanim, and E. H. Cordes, J. Phys. Chem., 73, 1898 (1969).

16. C. Bellorin, Trabajo especial de grado. Esc. Quimica. Fac. Ciencias, Universidad Central de Venezuela (1972).

17. A. Armas, H. Clemente, J. Coronel, F. Creazzola, A. Cuenca, J. Francis, A. Malpica, D. Quintero, R. Romero, J. Salazar, N. Sanchez, R. Von Bergen, J. Baumrucker, M. Calzadilla, and E. H. Cordes, J. Org. Chem., 37, 875 (1972).

18. A. Alam, Trabajo Especial de Grado Esc. Quimica Fac. Ciencias Universidad Central de Venezuela (1972).

19. C. A. Bunton, E. J. Fendler, L. Sepulveda, and K. -U. Yang, J. Amer. Chem. Soc., 90, 5512 (1968).

20. J. Baumrucker, M. Calzadilla, M. Centeno, G. Lehrmann, P. Lindquist, D. Dunham, M. Price, B. Sears, and E. H. Cordes, J. Phys. Chem., 74, 1152 (1970).

21. H. G. Bull, K. Koehler, T. C. Pletcher, J. J. Ortiz, and E. H. Cordes, J. Amer. Chem. Soc., 93, 3002 (1971).

22. J. Baumrucker and M. Calzadilla, unpublished results.

23. T. H. Fife and L. H. Brod, J. Amer. Chem. Soc., 92, 1681 (1970).

24. E. F. J. Duynstee and E. Grunwald, J. Amer. Chem. Soc., 81, 4540, 4542 (1969).

25. J. Albrizzio, J. Archila, T. Rodulfo, and E. H. Cordes, J. Org. Chem., 37, 871 (1972).

26. L. R. Romsted and E. H. Cordes, J. Amer. Chem. Soc., 90, 4404 (1968).

MICELLAR EFFECTS IN STEADY-STATE RADIATION INDUCED REACTIONS

J. H. Fendler, G. W. Bogan, E. J. Fendler,
G. A. Infante and P. Jirathana

Department of Chemistry
Texas A & M University
College Station, Texas 77843

The primary chemical result of the irradiation of water after the deposition of energy is the formation of the following species:

$$H_2O \xrightarrow{\sim\sim\sim} e^-_{aq} + \cdot H + \cdot OH + H_2 + H_2O_2 + H_3O^+ \quad (1)$$

With gamma rays the primary products given in reaction 1 are formed inhomogeneously in small widely spaced clusters in what has been known as spurs within a time scale of 10^{-10} to 10^{-8} second.[1] After the spur is formed, these entities diffuse into the bulk of the solvent or competitively react with each other and with the solvent to form H_2 and H_2O_2 or to reform water. Using scavengers the primary yields have been determined to be: $G_{e^-_{aq}} = 2.8 \pm 0.1$, $G_{\cdot H} = 0.6 \pm 0.1$, $G_{OH} = 2.6 \pm 0.1$, $G_{H_2} = 0.45$, and $G_{H_2O_2} = 0.71$.[2] A knowledge and appreciation of the rate constants for the primary chemical species with each other and with dissolved solutes[3] are commonly and profitably employed to simplify aqueous radiation induced reactions. Table I lists rate constants for some of the most frequently used reactions.

The system can be simplified to contain primarily e^-_{aq} by adding methanol to scavenge $\dot{O}H$ and $\cdot H$ and adjusting the pH to ca. 10 to scavenge H_3O^+ which could otherwise react rapidly with e^-_{aq}. In order to investigate the reactions of $\cdot H$ with solutes, the triply distilled water is made acidic to convert e^-_{aq} to $\cdot H$ and is saturated with hydrogen

Table I

Selected Rate Constants for the Primary Species in Water[a]

Reaction	Rate Constant,[b] $M^{-1} sec^{-1}$	pH
$e^-_{aq} + H_3O^+ \rightarrow \cdot H + H_2O$	$(2.07 \pm 0.08) \times 10^{10}$	2.1-4.3
$e^-_{aq} + e^-_{aq} \rightarrow H_2 + 2OH^-$	$(0.9 \pm 0.15) \times 10^{10[c]}$	10.9
$e^-_{aq} + H_2O_2 \rightarrow \cdot OH + OH^-$	$(1.23 \pm 0.14) \times 10^{10}$	7
$e^-_{aq} + \cdot H \rightarrow H_2 + OH^-$	$(2.5 \pm 0.6) \times 10^{10}$	10.5
$e^-_{aq} + \cdot OH \rightarrow OH^-$	$(3.0 \pm 0.7) \times 10^{10}$	10.5
$e^-_{aq} + \cdot O^- + 2OH^-$	$(2.2 \pm 0.6) \times 10^{10}$	13
$e^-_{aq} + H_2O \rightarrow \cdot H + OH^-$	16.0 ± 1.0	8.4
$e^-_{aq} + O_2 \rightarrow \cdot O_2^-$	$(1.88 \pm 0.2) \times 10^{10}$	7
$e^-_{aq} + N_2O \rightarrow N_2 + \cdot OH + OH^-$	$(8.67 \pm 0.6) \times 10^9$	7
$\cdot H + \cdot H \rightarrow H_2$	$1.5 \times 10^{10[c]}$	0.1-1.0
$\cdot H + OH^- \rightarrow e^-_{aq} + H_2O$	1.8×10^7	11-13
$\cdot H + \cdot OH \rightarrow H_2O$	$(0.7-3.2) \times 10^{10}$	3
$\cdot H + O_2 \rightarrow \cdot HO_2$	2.6×10^{10}	0.4-3.0
$\cdot H + H_2O_2 \rightarrow \cdot OH + H_2O$	$(9.0 \pm 1) \times 10^7$	2.1
$\cdot H + H_3O^+ \rightarrow H_2^+ + H_2O$	$2.6 \times 10^{3[d]}$	3.5-11
$\cdot H + N_2O \rightarrow N_2 + \cdot OH$	$\sim 1.2 \times 10^{4[d]}$	3.5-11
$\cdot OH + \cdot OH \rightarrow H_2O_2$	$5 \times 10^{9[c]}$	7
$\cdot OH + OH^- \rightarrow \cdot O^- + H_2O$	$3.6 \times 10^{8[e]}$	-
$\cdot OH + H_2O_2 \rightarrow H_2O + \cdot HO_2$	4.5×10^7	7
$\cdot OH + H_2 \rightarrow \cdot H + H_2O$	$(6.0 \pm 2.0) \times 10^7$	7
$\cdot OH + CH_3OH \rightarrow \cdot CH_2OH + H_2O$	4.8×10^8	7
$H_3O^+ + OH^- \rightarrow 2H_2O$	$1.43 \times 10^{11[f]}$	7

a) References 3 and 4. b) Determined by pulse radiolysis unless stated otherwise. c) Rate constant, k, defined by $d(X)/dt = k(X)^2$, where $X = e^-_{aq}$, $\cdot H$, or $\cdot OH$. d) Determined photochemically. e) Determined by competition kinetics. f) Determined by T-jump technique.

to scavenge $\dot{O}H$. Hydroxyl radical reactions are studied in the presence of nitrous oxide since it not only eliminates e^-_{aq} but doubles the amount of hydroxyl radical in the system. The state of the art of current day radiation chemistry therefore allows the design of meaningful experiments involving either the primary reducing species e^-_{aq}, which can be considered to be the simplest nucleophile, or the oxidizing species $\dot{O}H$, which has electrophilic properties. Additionally it is possible to investigate reactions of radiation induced halide ion radicals, $X_2^{-\cdot}$. $Cl_2^{-\cdot}$ radical can be generated, for example, by irradiating aqueous air saturated acidic sodium chloride solutions. Under these conditions, e^-_{aq} and $\cdot H$ react with oxygen (Table I) and $Cl_2^{-\cdot}$ is formed by:

$$\dot{O}H + Cl^- + H_3O^+ \rightarrow Cl\cdot + 2H_2O$$

and

$$Cl\cdot + Cl^- \rightarrow Cl_2^{-\cdot}$$

Mechanistic investigations of a number of radiation-induced reactions in aqueous systems have been prompted to a large extent by their relevance to biological processes <u>in vitro</u> and <u>in vivo</u>.[1] However, the validity of applying the results of radiation chemistry in <u>dilute aqueous solutions</u> to the more complex biological systems may be questioned. Relatively high local concentrations of solute in these systems might result, at least in part, in <u>direct</u> solute interactions with high-energy particles. Furthermore, naturally occurring macromolecules often exist in self-associating systems of aggregates and in several tertiary structures or conformations which may have profound effects on the reactivity of these macromolecules. Reaction of e^-_{aq} with ribonuclease, for example, was found to correspond to the reversible unfolding of the molecule as a function of increasing temperature.[5] Complex formation also affects the rates of radiation induced reactions; the reactivity of e^-_{aq} with cationic methylene blue decreases approximately eight-fold when the dye is bound to heparin.[6] The detailed radiation chemistry of proteins and other macromolecules is inevitably rather complex.

We have initiated systematic investigations using micellar systems to provide simple model environments for radiation-induced radical reactions. Micellar systems have several advantages in such investigations over the more complex systems of biological macromolecules. The physical properties of the aggregate or micelle, such as the critical micelle concentration (CMC), aggregation number, surface charge and substrate solubility, are known or can be determined.[7,8] Also, the CMC generally occurs in dilute solution, 10^{-2}-10^{-4} M, thereby permitting investigation of radiation-induced reactions in aqueous solution in the absence of direct radiation effects and scavenging from the spurs. Additionally, micelle-substrate systems can be designed which involve specific electrostatic, hydrophobic or neighboring group interactions.

One surfactant of each charge type has been utilized in our initial investigations: <u>cationic</u> hexadecyltrimethylammonium bromide (CTAB), <u>anionic</u> sodium dodecyl sulfate (NaLS), and <u>neutral</u> polyoxyethylene(15) nonylphenol (Igepal CO-730). The choice of these surfactants was governed by the availability of reliable physical data, such as the critical micelle concentration,[8] aggregation number, viscosity and substrate solubility, and by kinetic and thermodynamic data on nonradiation-induced ionic and radical reactions in aqueous micellar solutions.[7,8]

EXPERIMENTAL TECHNIQUES[9]

Our experimental approach involves determination of rate constants and products. Pulse radiolysis has been utilized to determine the reactivities of e^-_{aq}, $\dot{O}H$, and $Cl_2^{\overline{\cdot}}$ with surfactants in aqueous solutions both above and below their critical micelle concentrations. Since e^-_{aq} and $Cl_2^{\overline{\cdot}}$ have broad absorbances centered at 720 and 340 nm, respectively, reactions of these species can be monitored directly, at least in principle, by following the rate of absorbance decrease at given scavenger concentration. Conversely, the lack of a suitable absorbance for $\dot{O}H$, necessitates the use of the thiocyanate competition technique.[14,15] Thiocyanate anion reacts rapidly with $\dot{O}H$ to produce a transient absorbing species, $(SCN)_2^{\overline{\cdot}}$, with a maximum at 480 nm:

$$\dot{O}H + SCN^- \xrightarrow{k_1} \text{absorbing species} \left((SCN)_2^{\overline{\cdot}}\right) \qquad (2)$$

If the competitor, S (the surfactant, for example) does not produce a transient species which absorbs at this wavelength:

$$\dot{O}H + S \xrightarrow{k_2} \text{non-absorbing species} \quad (3)$$

then:

$$\frac{A_o}{A} = 1 + \frac{k_2[SCN^-]}{k_1[S]} \quad (4)$$

where A_o and A are the absorbances in the absence and presence of S, respectively. Values for the rate constant k_1 allow calculation of those for k_2.

Rate constants for the reactions of hydrogen atoms have been determined by steady-state competition methods using isopropanol-d_7 as the competitor and acetone to scavenge e_{aq}^- [16]:

$$\frac{G°(HD) - G(HD)}{G(HD)} = \frac{k_{S + H\cdot}[S]}{k_{\text{isopropanol-}d_7 + \cdot H}[\text{isopropanol-}d_7]} \quad (5)$$

where $G°(HD)$ and $G(HD)$ are the HD yields in the presence and absence of surfactants, respectively. Once again, values for $k_{\text{isopropanol-}d_7 + \cdot H}$ allow the calculation of $k_{S + H\cdot}$.

Due caution needs to be exercised, of course, to establish that the competitor, thiocyanate ion or isopropanol-d_7, does not react with the radiolytic species differentially in the monomeric and in the micellar environments. For all the competitors used, this has been established.

Subsequent to determining the rate constants for the reactions of e_{aq}^-, $\dot{O}H$, $\cdot H$ and Cl_2^- with the surfactants below and above their critical micelle concentrations, reactions of these species with substrates solubilized by the micelle could be investigated. Once again, both direct and competitive kinetics have been utilized.

In the irradiation of aqueous halogen-substituted carboxylic acids we have analyzed for halide ions by ion-selective electrodes and used gas-liquid-partition chromatography to determine yields of other products at different doses and solute concentrations.

Our most detailed study on the effects of micelles on radiolytic products involves thymine. The products formed on irradiation of carbon-14 labelled thymine have been separated by descending paper chromatography using two different solvent systems, and determined quantitatively by means of a radiochromatographic scanner. The products have been identified by their R_F values (Table II).

Irradiations were carried out using triply distilled water and freshly annealed glassware. The surfactants were purified according to established procedures.[8] No minima in plots of the surface tension vs. surfactant concentration were observed, indicating the absence of impurities. The purity of the surfactants was also confirmed by the absence of observable impurities in their infrared and ^{1}H nmr spectra. All scavengers were chromatographically pure.

Table II

R_F Values[a]

	CH_3CN:Aq.Na_2HPO_4-NaH_2PO_4 (85:15; v/v)		BuOH:H_2O (86:14; v/v)	
Thymine dimer	0.02	(0.02)	0.02	(0.02)
trans-5-Hydroperoxy-6-hydroxydihydrothymine	0.29	(0.29)	0.15	(0.15)
cis-Thymine glycol	0.35	(0.35)	0.21	(0.21)
trans-Thymine glycol	0.41	(0.40)	0.27	(0.27)
Thymine	0.50	(0.50)	0.50	(0.50)
Formylpyruvylurea	0.70	(0.70)	0.56	(0.55)
5-Hydroxy-5-methyl-barbituric acid	0.56	(0.56)	0.30	(0.29)

a) Values in parenthesis are from reference 17.

REACTIVITY OF e^-_{aq}, $\dot{O}H$, ·H AND Cl_2^- WITH MICELLE FORMING SURFACTANTS[10-13]

Table III summarizes the available rate constants for the reactions of the radicals with surfactants below and above their critical micelle concentrations.

As expected,[18] the reactivity of hydrocarbon-type surfactants with the hydrated electron is very low. $\dot{O}H$, $\dot{H}$ and Cl_2^- are more reactive toward monomeric surfactants by factors of 10^2-10^3 than hydrated electrons. Cl_2^- is much less reactive, however, than $\dot{O}H$, as has been observed for several organic reactions.[19]

Table III
Rate Constants for the Reactions of e^-_{aq}, $\dot{O}H$, H· and Cl_2^- with Micelle Forming Surfactants, S

	k, M^{-1} sec^{-1}			
	e^-_{aq} + S	$\dot{O}H$ + S	H· + S	Cl_2^- + S
CTAB				
[S] < CMC		1.04×10^{10}	1.6×10^8	
[S] > CMC	$\leq 9 \times 10^5$	2.1×10^9	1.6×10^8	
CTACl				
[S] < CMC				8.6×10^6
[S] > CMC				1.3×10^6
NaLS				
[S] < CMC		7.6×10^9	1.2×10^8	2.6×10^6
[S] > CMC	$\leq 2 \times 10^5$	5.0×10^8	1.2×10^8	2.4×10^4
Igepal CO-730				
[S] < CMC		1.1×10^{10}	2.1×10^9	2.4×10^9
[S] > CMC	$\leq 1 \times 10^6$	1.7×10^9	4.9×10^8	3.6×10^8

Rate constants for the reactions of $\dot{O}H$ and $Cl_2^{\overline{\cdot}}$ with micellar surfactants are smaller above than below their critical micelle concentrations. Such behavior is best illustrated by the directly determined rate constants for $Cl_2^{\overline{\cdot}}$ with Igepal CO-730 as a function of surfactant concentration (Figure 1). These results bring into question the validity of extending rate constant determinations of $\dot{O}H$ and $Cl_2^{\overline{\cdot}}$ reactions from aqueous solutions to biological macromolecular systems.

The reactivities of CTAB and NaLS toward $\dot{H}$ are apparently not altered by micelle formation. Igepal which does show phase-dependent behavior contains a phenyl group which is deeply buried in the micelle whereas in the other micellar surfactants, the reactive sites near the surface are much like those in the core of the micelle. Clearly then, the changes in stoichiometry occurring above the CMC do not alone determine reactivity of the micellar phase.

Figure 1

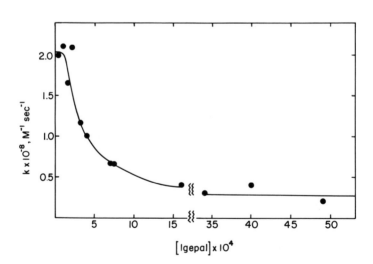

Rate constants for the reaction of $Cl_2^{\overline{\cdot}}$ with Igepal CO-730 as a function of surfactant concentration

REACTIVITY OF SOLUBILIZED BENZENE WITH e^-_{aq} AND $\dot{O}H$ IN MICELLAR SURFACTANT SOLUTIONS[12]

The rate constant for electron addition to benzene as a function of NaLS concentration reaches a minimum value which is three-fold smaller than that in pure water (Figure 2). With higher concentrations of benzene the rate inhibition becomes greater.[10] Using absorption[20] and proton magnetic resonance spectroscopy[21] the primary solubilization site of benzene was demonstrated to be the micellar interior. The observed rate inhibition implies either that the rate of electron attachment is different in the two phases or that the penetration of e^-_{aq} to the site of the solubilized benzene is hindered by the outer structure of the micelle.

Rate constants for electron addition to benzene is enhanced by micellar CTAB (Figure 3). Since benzene is solubilized at the CTAB-water interface,[21,22] these results are explicable in terms of electrostatic interactions between the π-electron system of the benzene molecule and the net positive charge on the CTAB micelle surface which are likely to render benzene more susceptible to nucleophilic attack by the electron.

Figure 2

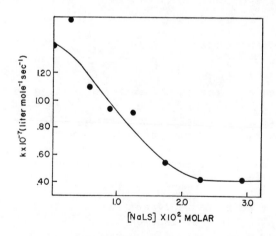

Rate constants for the reaction of e^-_{aq} with benzene as a function of NaLS concentration

Figure 3

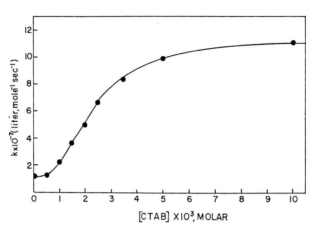

Rate constants for the reaction of e_{aq}^{-} with benzene as a function of CTAB concentration

Making the usual assumptions,[8] equation 6 allows the estimation of the binding constant, K, between benzene and the micellar surfactant.

$$\frac{k_\psi - k_o}{k_m - k_\psi} = \frac{K}{N}(C_D - CMC) \qquad (6)$$

where k_ψ, k_o, and k_m are rate constants at a given surfactant concentration, in water, and in the micellar phase, C_D is the stoichiometric surfactant concentration, N is the aggregation number, and CMC is the critical micelle concentration. K values of ca. 10^4 M^{-1} have been obtained for both CTAB and NaLS in fair agreement with those derived from solubility measurements.

All micellar surfactants retard the rate of addition of ȮH to benzene. The rate retardation of the ȮH attack on benzene solubilized in NaLS also can be rationalized by the inaccessibility of the substrate and consequently the catalyses in CTAB solutions are consistent with the different polarities of $\dot{O}H$ and e_{aq}^-. Table IV summarizes the available rate constants for the reaction of e_{aq}^- and $\dot{O}H$ with benzene in the different micellar surfactants.

Table IV

Rate Constants for the Reactions of e_{aq}^- and $\dot{O}H$ with Benzene in Water and in the Presence of Micellar Surfactants[a]

Surfactant	$k_{(e_{aq}^- + C_6H_6)}$, $M^{-1} sec^{-1}$	$k_{(\dot{O}H + C_6H_6)}$, $M^{-1} sec^{-1}$
None[b]	1.3×10^7	8.2×10^9
CTAB	11.0×10^7	2.6×10^9
NaLS	0.4×10^7	3.0×10^9
Igepal CO-730	0.6×10^7	2.6×10^9

a) In all cases the surfactant concentration exceeded the CMC by no less than a factor of 5 with the exception of the NaLS system in the reaction with $\dot{O}H$.

b) Rate constants in pure water.

RADIOLYSIS OF THYMINE

Radiation-induced reactions of pyrimidine bases in aqueous solutions involve deamination, ring opening and the formation of several products.[1] The radiation chemistry of this system has been elucidated from determinations of the base destruction yield G(-pyrimidine), spectrophotometrically, as well as from product analyses. We initiated studies of micellar effects on the radiolysis of thymine since the radiolytic decomposition of this compound has been well established.[17,23-24] Figure 4 summarizes the postulated mechanism.[17,24]

Figure 4

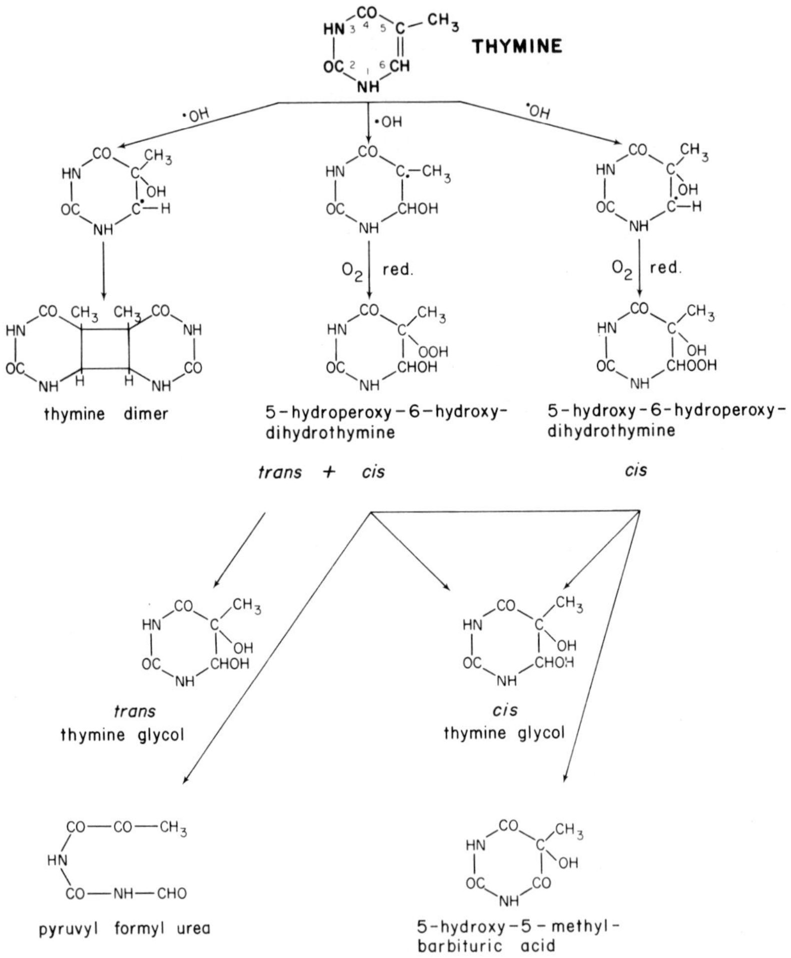

Suggested mechanism for thymine radiolysis in aqueous air saturated solutions

The first step in the radiolytic decomposition has been established by means of pulse radiolysis and esr spectroscopy to be the addition of $\dot{O}H$, $\cdot H$ or e_{aq}^- across the 5-6 double bond.[1] The remainder of the reaction scheme is based solely on the analysis of products.

Using C-14 labelled thymine we have quantitatively determined the yields for the thymine-dimer, for the peroxy compound, for the *cis*- and *trans*-glycols, for formylpyruvylurea, and for 5-hydroxy-5-methylbarbituric acid as the radiolytic products. Figure 5 illustrates typical radiochromatographic charts at a given dose (100,000 R). G-values have been calculated from linear yield-dose plots containing at least 5 points (Figure 6). The results are summarized in Table V.

It should be pointed out that an almost complete material balance is obtained; G(-thymine) $\cong$ 0.9 G(products). Surfactants decrease the base destruction, G(-thymine), values. More significant, however, are their effects on the product yields particularly those on the peroxy compound and the dimer. The decreased yield of the *cis*- and *trans*-thymine glycols, formylpyruvylurea and 5-hydroxy-5-methylbarbituric acid *below the CMC* is accountable in terms of competition between thymine and the monomeric surfactants for $\dot{O}H$. *Above the CMC*, decrease in the product yields are greater than expected on this competition alone and are specific indicating pronounced micellar effects.

Since chloride ions have been found to affect the base destruction yield of pyrimidines and since radiation protection and sensitization by chloride ion have important biological consequences,[19] we have investigated the radiation induced reactions of $Cl_2^{\cdot -}$ with thymine in the presence of micelle forming surfactants both by pulse and steady-state radiolysis.[13] Micelles do not appreciably affect the first step, but dramatically interfere with the subsequent chain reactions.

Micellar effects on radiation induced radical reactions with thymine are in agreement with substrate solubilization studies in that thymine is not solubilized by micelles and hence there is no effect on the rates of reactions involving it. Interaction of the subsequently formed species with micelles are, therefore, responsible for the observed micellar effects.

Figure 5

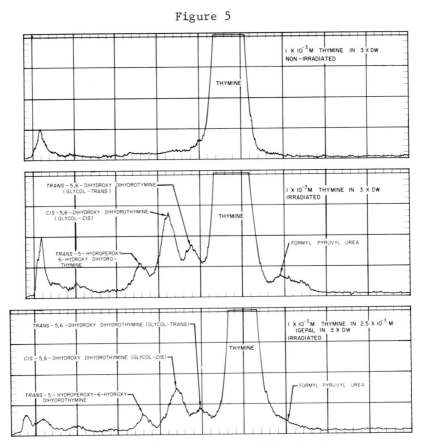

Typical radiochromatographic charts

Figure 6

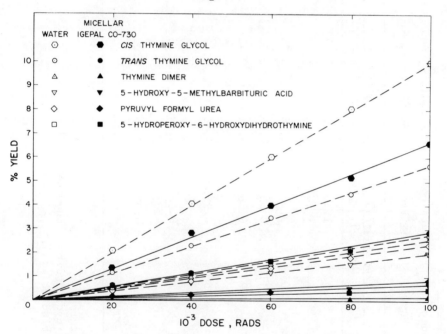

Typical yield dose plots for product formation in irradiated thymine in aqueous air saturated solution in the absence (solid line) and presence of micellar Igepal CO-730 (broken line)

Table V

G-Values for the Radiolytic Products Formed in Irradiated Aqueous Air-Saturated Thymine in the Presence of Surfactants, S

	Water	Igepal CO-730		NaLS		CTAC1	
		[S] < CMC	[S] > CMC	[S] < CMC	[S] > CMC	[S] < CMC	[S] > CMC
G(-Thymine)[a]	2.71	1.92	1.20	2.05	1.12	2.27	1.62
G(-Thymine)[b]	2.71	2.04	1.23	1.99	1.05	2.28	1.55
G(Thymine dimer)	0.22	0.05	0.02	0.16	0.08	0.13	0.03
G(Peroxy)[c]	0.33	0.11	0.03	0.04	0.03	0.05	0.04
G(Cis-thymine glycol)	0.97	0.91	0.58	0.71	0.42	0.94	0.66
G(Trans-thymine glycol)	0.54	0.50	0.25	0.40	0.26	0.51	0.40
G(Formylpyruvyl-urea)	0.23	0.21	0.08	0.18	0.13	0.17	0.15
G(Barbituric acid)[d]	0.16	0.13	0.06	0.05	0.02	0.05	0.02
G(Product X)				0.14		0.25	0.15

a) Determined by spectra.
b) Determined by chromatography.
c) Trans-5-hydroperoxy-6-hydroxydihydrothymine.
d) 5-Hydroxy-5-methylbarbituric acid.

CONCLUSION

Formation of association colloids and substrate solubilization therein substantially affect the rates of radiation-induced reactions. These effects are specific and are dependent on both the nature of the substrate and the charge type of the micelle. The micellar rate enhancements or retardations generally obey Michaelis-Menten kinetics and the magnitude of substrate-micelle binding constants is of the order of 10^4 M^{-1}. Although micellar systems are far from perfect models for complex radiation biological systems, they provide a better approximation than pure water.

We are currently investigating radiation-induced reactions of nucleotides, nucleosides and other bio-organic molecules in micellar systems. Pulse radiolytic determinations of the reaction rates and analyses of the product yields from steady-state irradiation experiments coupled with ^1H and ^{19}F nmr spectroscopic and uv spectrophotometric investigations of the environment of the substrates in the micellar phase will allow elucidation of the mechanisms of micellar catalysis in these systems. It is anticipated that the results of these studies will ultimately lead to the design of radiation biological model systems involving highly specific interactions.

ACKNOWLEDGMENTS

Support for this work from the U. S. Atomic Energy Commission and from the Robert A. Welch Foundation is gratefully acknowledged. Dr. E. J. Fendler is a National Institutes of Health Research Career Development Awardee.

REFERENCES

1. E. J. Fendler and J. H. Fendler, Prog. Phys. Org. Chem., 7, 229 (1970) and references cited therein.

2. G(X) and G(-X) refer to the number of molecules of product X formed or decomposed, respectively, on irradiation per 100 eV of absorbed energy. The notation G_x is used to designate the primary radical yield and

distinguishes it from the observed yield, G(X). For example, $G(H_2)$ is rarely equal to G_{H_2} since the hydrogen is formed not only by primary recombination of e_{aq}^-, ·H and ·OH but also via reactions of these species with the solutes or other species formed during radiolysis.

3. M. Anbar and P. Neta, Int. J. Appl. Radiat. Isotopes, 18, 493 (1967).

4. M. Anbar, "Selected Specific Rates of Reactions in Aqueous Solution. I. Hydrated Electron. II. Hydrogen Atom. III. Hydroxyl Radical. IV. Perhydroxyl Radical," to be published. (For information write Radiation Chemistry Data Center, Radiation Laboratory, Univ. of Notre Dame, Notre Dame, Ind. 46556).

5. R. Braams and M. Ebert, in "Radiation Chemistry. I" (Ed. R. F. Gould), "Advances in Chem. Ser., 81, 464 (1968).

6. L. I. Grossweiner, in "Radiation Chemistry. I" (Ed. R. F. Gould), "Advances in Chem. Ser., 81, 309 (1968).

7. E. H. Cordes and R. B. Dunlap, Accounts Chem. Res., 2, 329 (1969).

8. E. J. Fendler and J. H. Fendler, Advan. Phys. Org. Chem., 8, 271 (1970).

9. More details can be found in references 1 and 10-13.

10. J. H. Fendler and L. K. Patterson, J. Phys. Chem., 74, 4608 (1970).

11. L. K. Patterson, K. M. Bansal and J. H. Fendler, Chem. Commun., 152 (1971).

12. K. M. Bansal, L. K. Patterson, E. J. Fendler and J. H. Fendler, Int. J. Radiat. Phys. Chem., 3, 321 (1971).

13. J. H. Fendler, E. J. Fendler, G. Bogan, L. K. Patterson and K. M. Bansal, J.C.S. Chem. Commun., 14 (1972).

14. G. E. Adams, J. W. Boag, J. Current and B. D. Michael, in "Pulse Radiolysis," Academic Press, 1965, p. 117.

15. C. L. Greenstock, J. W. Hunt and M. Ng, Trans. Faraday Soc., 65, 3279 (1969).

16. P. Neta, G. R. Holdren and R. H. Schuler, J. Phys. Chem., 75, 449 (1971).

17. R. Teoule, D.Sc. Thesis, University of Lyons, 1970, No. 667.

18. M. Anbar, Advan. Phys. Org. Chem., 7, 115 (1969); E. J. Hart and M. Anbar, "The Hydrated Electron," Wiley-Interscience, New York, 1970.

19. J. F. Ward and I. Kuo, Advan. Chem. Ser., 81, 368 (1968).

20. S. J. Rehfeld, J. Phys. Chem., 74, 117 (1970).

21. E. J. Fendler, C. L. Day and J. H. Fendler, J. Phys. Chem., 76, 1460 (1972).

22. J. C. Eriksson and G. Gillberg, Acta Chem. Scand., 20, 2019 (1966).

23. B.-S. Hahn and S. Y. Wang, J. Amer. Chem. Soc., 94, 4764 (1972).

24. R. Teoule and J. Cadet, Chem. Commun., 1269 (1971).

ELECTROLYTE EFFECTS ON MICELLAR CATALYSIS

C. A. Bunton

Department of Chemistry
University of California
Santa Barbara, California 93106

In classifying chemical reactions, it is convenient to define the mechanism in terms of the molecularity of the slow, rate limiting, step. If only the substrate is undergoing covalency changes in this step, the reaction is considered to be unimolecular, i.e., it follows a dissociative mechanism, but if both substrate and an external reagent undergo covalency changes in this step, the reaction is considered to be bimolecular, i.e., it follows an associative mechanism. This nomenclature system is used extensively, and has been discussed in detail in numerous monographs [1,2,3].

I will use phosphate ester hydrolysis to illustrate these mechanistic principles of molecularity; because micellar effects upon these hydrolyses will be considered in this discussion. Two unimolecular mechanisms of phosphate ester hydrolysis with phosphorus—oxygen bond cleavage will be discussed; one involving spontaneous elimination of a metaphosphate anion from the monoanion of a phosphate monoester [4,5]:

$$RO - \underset{\underset{O}{\|}}{\overset{\overset{H-O}{|}}{P}} - O^- \xrightarrow{\text{slow}} ROH + PO_3^- \xrightarrow[\text{fast}]{H_2O} H_2PO_4^-$$

and the other, a similar spontaneous elimination from a dianion [6,7]:

$$RO-PO_3^{2-} \xrightarrow{slow} RO^- + PO_3^- \xrightarrow[fast]{H_2O} H_2PO_4^-$$

Each of these mechanisms can be observed by making the appropriate choice of substrate structure and reaction medium, e.g., pH or polarity of the solvent.

In the bimolecular mechanisms, a nucleophile attacks either the phosphoryl or the alkyl or aryl group [4,5,8], and two typical examples of attack on the phosphoryl group are:

$$NO_2-C_6H_4-O-\overset{O}{\underset{}{P}}(OPh)_2 \xrightarrow{X^-} NO_2-C_6H_4-O^- + X-\overset{O}{\underset{}{P}}(OPh)_2 \qquad \text{Ia}$$

$$NO_2-C_6H_3(NO_2)-O-\overset{O}{\underset{O^-}{P}}-OR \xrightarrow{X^-} NO_2-C_6H_3(NO_2)-O^- + X-\overset{O}{\underset{O^-}{P}}-OR \qquad \text{Ib}$$

$$X^- = OH^-, F^-, RO-PO_3^{2-}$$

The substitution product (Ia,b) may decompose further, depending upon the nature of X and the reaction medium. For example if X is fluoride, the intermediate phosphoryl fluoride will hydrolyze readily [8].

There are other mechanistically important reactions of phosphate esters, for example those involving acid catalysis, or carbon-oxygen bond cleavage [4,5], which I will not discuss here.

We would expect that attack of an anion upon a phosphate ester would be catalyzed by cationic and inhibited by anionic micelles, as is found. (For general discussions of micellar catalysis and inhibition, see refs. 9-11). For example cationic micelles of cetyl trimethylammonium bromide (CTABr, $C_{16}H_{33}NMe_3Br$) take up a nonionic phosphate ester into an environment in which it is exposed to attack by hydroxide or fluoride anions, whereas incorporation into anionic micelles of, for example sodium lauryl sulfate (NaLS, $C_{12}H_{25}SO_4Na$) protects the ester from nucleophilic attack by external anions [12]. Some examples of these kinetic effects are given in Table I.

TABLE I
MICELLAR CATALYSIS AND INHIBITION ON SOME UNI- AND BIMOLECULAR REACTIONS

Reaction	CTABr	NaLS	Igepal
Unimolecular			
2-NO_2,4-O_2N-C$_6$H$_3$-OPO_3^{2-} → 2-NO_2,4-O_2N-C$_6$H$_3$-O^- + PO_3^-	increase ≈ 25-fold	none	none
benzofuran-CO_2^- (O_2N, CN) → O_2N-C$_6$H$_3$(CN)-O^- + CO_2	increase ≈ 100-fold	none	slight increase
PhCH(CN)CO_2^- → PhC̄HCN + CO_2	increase ≈ 660-fold		
PhCH(Br)$CH_2CO_2^-$ → PhC̊HCH$_2$CO$_2^-$ + Br^-	decrease		
Bimolecular			
O_2N-C$_6$H$_4$-O-P(=O)(OPh)$_2$ + OH^- → O_2N-C$_6$H$_4$-O^- + (PhO)$_2$P(=O)OH	increase	decrease	decrease
O_2N-C$_6$H$_4$-O-P(=O)(OPh)$_2$ + F^- → O_2N-C$_6$H$_4$-O^- + (PhO)$_2$P(=O)F	increase	decrease	decrease

The micelles are therefore behaving like a microenvironment with the reaction occurring in them or on their surface, just as is observed in heterogeneous catalysis, or in catalysis by polyelectrolytes [9] or liquid crystals [13] or in enzymic catalysis [14]. It is generally assumed that micelles are approximately spherical, provided that the surfactant concentration is not high [10,15]; but that at high surfactant concentrations, and especially in the presence of some added electrolytes, the micelles cease to be spherical; and that this tendency is especially pronounced with micelles of nonionic surfactants, where there is no electrostatic repulsion between ionic head groups [16].

It is also readily understandable that added electrolytes will tend to exclude anionic reagents from the Stern layer of an ionic micelle and so reduce or even eliminate the micellar catalysis, and these salt inhibitions of micellar catalysis have been observed for many bimolecular reactions [10,11]. They show certain common features; (i) salt effects are specific, and depend upon the nature of the ion which has the charge opposite to that of the micelle, and (ii) the effects are greatest for large, low charge density ions which interact most strongly with the counter ionic micelle. Examples of these electrolyte effects are the attack of hydroxide ion upon carboxylic esters catalyzed by cationic micelles where the anion order of inhibition is [17]:

$NO_3^- > Br^- > Cl^- > F^- >$ no salt;

for the corresponding reactions of the halonitrobenzenes [18] or triaryl phosphates [12], the order of inhibition is:

$OTos^- > NO_3^- > Br^- > Cl^- \sim CH_3SO_3^- > F^- >$ no salt;

for the acid hydrolysis of methyl orthobenzoate [20], the order of inhibition is:

$R_4N^+ > Cs^+ > Rb^+ > K^+ > Na^+ > Li^+ >$ no salt; c.f. [19]

for the rate of protonation of a vinyl phosphate [21], a protonation of an indicator in the presence of anionic micelles [22], the order of inhibition is:

$Me_4N^+ > Na^+ > Li^+$.

The salt order is essentially independent of the nature of the bimolecular reaction, and it appears to apply to reaction rates and equilibria.

This qualitative explanation of salt effects emphasizes competition between salt and reagent for the micelle and deemphasizes effects upon micellar structure although added salts increase micellar size, and therefore reduce the number of micelles at any given surfactant concentration [23].

Considering first micellar catalysis in the absence of large concentrations of added electrolytes, we would expect the rate constants of a micellar catalyzed reaction to increase to a plateau value when all the substrate is incorporated into the micelles, and this behavior is observed in a few reactions. One example which will be considered here is the unimolecular elimination of metaphosphate ion from the dianion of a dinitrophenyl phosphate dianion [24]. The kinetic form of the reaction catalyzed by cationic micelles can be treated quantitatively using the following scheme:

$$nD \rightleftharpoons D_n$$

$$D_n + S \underset{k_w}{\overset{K}{\rightleftharpoons}} SD_n \xrightarrow{k_m} \text{products}$$

(where S is the substrate and D the surfactant, or detergent, and D_n is the binding site for one substrate molecule)

This scheme with certain assumptions, leads to the following equation [10, 25]:

$$k_\psi = \frac{k_w + k_m K[D_n]}{1 + K[D_n]} \qquad (1)$$

and writing $[D_n] = (C_D - cmc)/N$, we can rearrange equation (1) into a reciprocal form which is akin to the Lineweaver-Burk treatment of enzyme kinetics [25]. The equation assumes that only one substrate molecule is taken up per micelle. This assumption is not necessarily correct because the micelle might have a number of binding sites, each of which could take up a substrate molecule. If this were the case, N would be smaller than the aggregation number of the micelle.

This simple kinetic treatment of micellar catalysis and inhibition predicts that there will be no kinetic effect at concentrations of the surfactant below the cmc, and this behavior is often found. However the treatment is only qualitative; in part because the reactant molecules can induce micellization and change the structure of the micelle, and submicellar aggregates could affect the reaction rate.

Micellar inhibition of bimolecular reactions can be treated satisfactorily using equation (1), but the equation is less satisfactory for micellar catalyzed bimolecular reactions [18]. However it has been applied satisfactorily to micellar catalyzed unimolecular reactions, where external reagents cause no complications [24,26]. The quantitative treatment can be extended to electrolyte inhibition [24], using the following scheme, where I is the inhibiting ion:

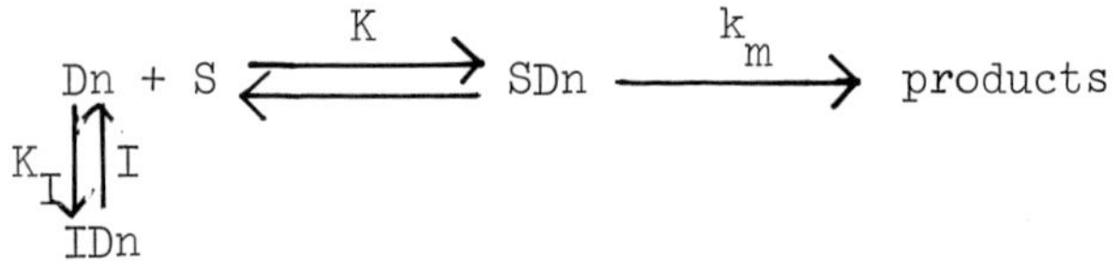

$$Dn + S \underset{}{\overset{K}{\rightleftarrows}} SDn \xrightarrow{k_m} products$$
$$K_I \updownarrow I$$
$$IDn$$

If we assume that the complex IDn is not a catalyst, and that the salt acts only by excluding the substrate from the micelle, we obtain equation (2):

TABLE II
SALT INHIBITION OF THE CTABr CATALYZED HYDROLYSIS OF THE 2,6-DINITROPHENYL PHOSPHATE DIANION[a]

Inhibiting Salt	K_IN/K	K_I
NaCl	0.01	7 (4)
CH_3SO_3Na	0.04	24
$C_6H_5OPO_3Na_2$	0.36	230
o-$C_6H_4(CO_2K)_2$	0.64	410 (370)
p-$C_6H_4(CO_2K)_2$	0.74	470
$C_6H_5CO_2Na$	1.26	810 (870)
NaOTos	4.4	2800

(a) For reaction in 3×10^{-3} M CTABr at 25.0°. The values of K_I in parentheses are for 2,4-dinitrophenyl phosphate dianion [24].

$$\frac{k_m - k_w}{k_\psi - k_w} = 1 + \frac{N}{K(C_D - cmc)} + \frac{K_I C_I N}{K(C_D - cmc)} \quad (2)$$

The second term is small compared with the third, so that for a given surfactant concentration, a plot of $(k_m - k_w)/(k_\psi - k_w)$ against C_I should be linear, as is found for the CTABr catalyzed hydrolyses of 2,4- and 2,6-dinitrophenyl phosphate dianions [24].

Using the treatment based on equation (2), we calculate $K_I N/K$ for the inhibition of the CTABr catalyzed hydrolysis of the 2,4- and 2,6-dinitrophenyl phosphate dianions by added salts (Table II). We cannot separate the salt effects upon K, K_I and N from the kinetic effects, although if we assume that the salt does not affect K and N, we can calculate K_I. The treatment also assumes that changes in the shape of the micelle do not of themselves affect the rate constant, km. This assumption appeared to be correct for the hydrolysis of the 2,4-dinitrophenyl phosphate dianions where changes in the shape of the CTABr micelle did not change km. In addition the calculation of K_I neglects the effect of the salts upon the cmc, and to this extent the K_I values quoted in Table II are too low. The values of the inhibition constants increase markedly with decreasing charge density or increasing hydrophobicity of the anion as is typical of electrolyte inhibition of micellar catalysis.

In all these kinetic treatments, we make some dubious assumptions; for example we neglect changes in micellar structure which may be brought about by addition of reagents or added electrolyte, but nonetheless the treatments appear to fit at least qualitatively. The salt effects shown in Table II are specific in that they depend upon the charge density of the inhibiting ion, and they provide a means of controlling the reaction in that we can stop a micellar catalyzed reaction simply by adding relatively small amounts of electrolyte, and in the same way, we can suppress micellar inhibition by adding electrolytes.

This pattern of electrolyte inhibition appeared to be general, but decarboxylation proved to be an exception. The spontaneous decarboxylation of the 6-nitrobenzisoxazole-3-carboxylate ion (II) is catalyzed (ca. 100-fold) to a plateau value by micelles of CTABr and 410-fold by

TABLE III
MICELLAR CATALYSIS OF DECARBOXYLATION[a]

Substrate	Surfactant	k relative
6-nitrobenzisoxazole-3-carboxylate	$C_{16}H_{33}\overset{+}{N}Me_3Br^-$	90
	$C_{12}H_{25}\overset{+}{N}Me_3Br^-$	65
	$\{C_{16}H_{33}\overset{+}{N}Me_2(CH_2)_3\}_2 2Br^-$	410
	$C_{12}H_{25}\overset{+}{N}Me_2CH_2CO_2^-$	180
	$C_{16}H_{33}\overset{+}{N}Me_2CH_2CH_2O^- (+OH^-)$	≈ 200
	$C_{16}H_{33}\overset{+}{N}Me_2CH_2CH_2OH$	85
$PhCHCO_2^-$ with CN	$C_{16}H_{33}\overset{+}{N}Me_3Br^-$	660
	$C_{12}H_{25}\overset{+}{N}Me_3Br^-$	300
	$C_{12}H_{25}\overset{+}{M}e_2CH_2CO_2^-$	700

(a) At 25° in water.

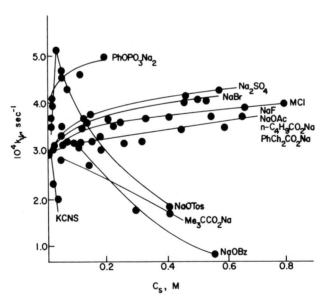

Figure 1. Effect of added salts upon the decarboxylation of 6-nitrobenzisoxazole-3-carboxylate ion in 2×10^{-2} M CTABr.

micelles of a dicationic surfactant (Table III and [27]).

This catalysis was not unexpected because in forming the transition state, the charge on the highly solvated carboxylate oxygens is delocalized into the ring so that a cationic micelle should stabilize the transition state relative to the initial state. Unexpectedly, the rate of the CTABr catalyzed decarboxylation is increased by a number of salts [28]. Some results for reaction in CTABr, in the plateau region, are shown in Figure 1. Some other

TABLE IV

EFFECT OF ADDED SOLUTES ON THE CTABr CATALYZED DECARBOXYLATION OF 6-NITROBENZISOXAZOLE-3-CARBOXYLATE ION[a]

Additive	10^2C additive	$10^4 k_\psi$, sec^{-1}
none		2.95
NaOTos	1.6	3.69
	3.4	5.09
	4.5	4.64
	7.0	4.47
	17.0	3.06
Na β-naphthalene sulfonate	0.55	3.50
	1.11	4.65
	1.66	2.50
$C_6H_5PO_4Na_2$	3.9	4.08
	14.4	4.37
	23.3	4.91
Na cholate	0.32	2.45
	0.64	1.80
estrone	0.45	3.43
	0.68	3.15
testosterone	0.13	2.50
	0.26	2.20
dioxane	179	2.65
	230	2.20
benzene	21	2.99
	45	3.08

(a) At 25.0° with 2×10^{-2} M CTABr [28].

examples of effects are given in Table IV. The rate enhancements by added ionic and nonionic solutes are relatively small (e.g., ~ 60%), but they contrast sharply with the usual salt inhibition. For several salts, all of which have anions which contain a phenyl group, the rate constant increases on addition of small amounts of the salt, and then decreases, but only a few salts, e.g., sodium cholate and trimethylacetate and potassium thiocyanate, show the typical salt inhibition at all concentrations. We were restricted in the salts which we could study; for example small amounts of sodium p-t-butylbenzene sulfonate increase the rate, but the solutions become so viscous that only low salt concentrations can be studied.

The micellar catalyzed decarboxylation of the anion of 2-cyano-2-phenylacetate ion (III) provides another example of this anomalous behavior of added salts [29]. This

$$\text{PhCHCO}_2^- \longrightarrow \text{PhCH}^- + \text{CO}_2 \xrightarrow{\text{fast}} \text{PhCH}_2\text{CN}$$
$$\quad\ \ |\ \ \qquad\qquad\qquad\quad |$$
$$\quad\ \text{CN} \qquad\qquad\qquad\ \ \text{CN}$$

III

reaction is catalyzed ca. 660-fold by micelles of CTABr (Table III), and salts which have relatively hydrophilic anions, e.g., Cl^-, SO_4^{2-}, HPO_4^{2-}, again increase the micellar catalysis (Table V). For this system, we could not use salts of aromatic acids because their ultraviolet absorption interferes with the rate measurements.

These results suggest that salts can perturb micellar

TABLE V
SALT EFFECTS UPON THE DECARBOXYLATION
OF 2-CYANOPHENYLACETATE ION IN CTABr[a]

Salt	C salt, M·	
	0.2	0.4
Na_2SO_4	1.84	2.22
NaCl	2.15	2.37
KCl	2.10	
KH_2PO_4		1.74
K_2HPO_4	1.75	2.17

(a) Values of $10^3 k_\psi$, sec^{-1} at 33.2° in 0.15 M CTABr; in the absence of salt, $k_\psi = 1.35 \times 10^{-3}$ sec^{-1} [29].

catalysis not only by competing with the substrate for the cationic micelle, but also by changing the catalytic power of the micelle, and that an increase in the catalytic power of the micelle might overcome the usual salt inhibition. In order to explain this behavior, we must consider the nature of micellar catalysis in some detail.

Incorporation of a substrate into a micelle stabilizes the initial state, relative to the substrate in water, and of itself should retard the reaction, and micellar catalysis means that micellar stabilization of the transition state more than offsets the stabilization of the initial state.

The increase of micellar catalysis of decarboxylation by many added hydrophilic salts suggests that partial neutralization of the micellar charge by added anions is an important factor, provided that the net micellar charge does not change so that the anionic substrate is no longer taken up by the micelle. In support of this hypothesis is the observation by Dr. Minch and Mr. Kamego that micelles of the zwitterionic surfactants (IV, V) are very good catalysts for the decarboxylation of 6-nitrobenzisoxazole-3-carboxylate ion. Some values for these rate enhancements

$$C_{12}H_{23}\overset{+}{N}Me_2CH_2CO_2^- \qquad C_{16}H_{33}\overset{+}{N}Me_2CH_2CH_2O^-$$

$$\text{IV} \qquad\qquad\qquad \text{V}$$

(k relative) of decarboxylation are given in Table III.

These observations show that a zwitterionic micelle with no net charge, and with the negatively charged groups closest to the micellar surface, will take up carboxylate ions from water by virtue of the hydrophobic interactions. Examples of catalysis of bimolecular reactions by micelles of zwitterions and liposomes of sonicated phospholipids have been studied by Professor Cordes and his coworkers [30]. However we found that neither lysolecithin nor sonicated α-lecithin catalyzed the decarboxylation of 6-nitrobenzisoxazole-3-carboxylate ion, suggesting that the nature of the micellar surface is of importance in catalysis of this decarboxylation.

Because of the unexpected kinetic behavior of sodium tosylate and of other salts in which the anion contained a phenyl group, we examined the interactions between micellized CTABr and several sodium arenesulfonates and benzoates. Most of our work was done using sodium tosylate, and I will focus my attention on its effects. At the same time, Professor Sepulveda and his coworkers at the University of Chile found that the viscosity of a solution of CTABr increases very markedly when sodium tosylate is added, and they observed a maximum viscosity when the concentrations of CTABr and tosylate ions were approximately equal, just as was found for the rate constant of decarboxylation of 6-nitrobenzisoxazole-3-carboxylate ion (Figure 2). These observations suggest that the micelle becomes rodlike when sodium tosylate is added to CTABr, but that the rodlike micelles either break up or revert towards spherical micelles when tosylate ion is in excess. Similar, but less well marked effects, were found with sodium benzene sulfonate, but salts with small hydrophilic anions had relatively small effects on the viscosity of CTABr. In transport experiments, Professor

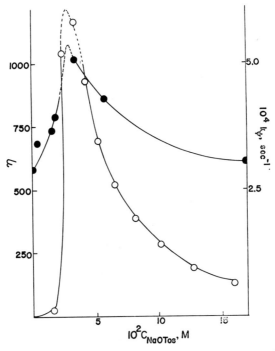

Figure 2. Relation between rate constant, k_ψ, of decarboxylation of 6-nitrobenzisoxazole-3-carboxylate ion and viscosity, for mixtures of CTABr-NaoTos. Solid point, k_ψ; open points, η.

Sepulveda found that when CTABr is in excess over sodium tosylate, the tosylate ion is transported to the cathode with the CTA^+, but when tosylate ion is in excess, the ammonium ion is transported to the anode.

There is extensive evidence that aromatic solutes have marked effects upon the shapes of ionic micelles; for example their initial addition to micelles of CTABr or cetylpyridinium chloride increases the lengths of the micelles, but in some cases an excess of the aromatic solute converts these large rodlike micelles into shorter, more compact micelles. Aliphatic hydrocarbons, e.g., methylcyclohexane have relatively little effect on micellar shape although like other hydrophobic solutes, they increase the aggregation numbers of cationic micelles, so that there appears to be some special interaction between aromatic solutes and cationic micelles [31]. Micellar effects on the reaction of the solvated electron with benzene have also been interpreted on the assumption that benzene is located close to the surface of a cationic micelle [32].

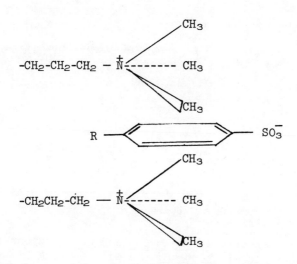

Figure 3. Suggested structure of the CTA^+ $OTos^-$ comicelle.

We have carried out NMR experiments at 100 MHz to examine the interactions between micellized CTABr and arene sulfonate and phenyl phosphate ions, and the best model for the comicelle is shown in Figure 3. The phenyl group is inserted into the micelle between the cationic heads so that the sulfonate group can still be solvated by water, and the electron rich phenyl group can interact with the cationic head groups of the surfactant. The viscosity measurements suggest that the surfactant chains are approximately parallel to each other in an elongated micelle. The repulsions between the cationic head groups are decreased, so that the micelle increases in size and becomes elongated. A brief summary of the NMR evidence follows. Relatively dilute solutions were used, so that it was necessary to use time averaging repetitive scans.

When arene sulfonate or benzoate or phenyl phosphate ion is added to CTABr, the ortho aromatic protons are shifted downfield relative to the meta, or para, protons.

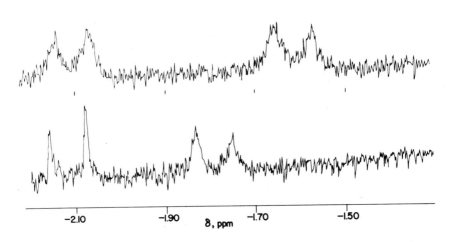

Figure 4. NMR spectra of the ortho and meta protons of 0.015 M NaOTos alone (lower spectrum) and with 0.024 M CTABr (upper spectrum). Chemical shifts are downfield from HOD signal.

An example of this behavior is shown in Figure 4 for a mixture of CTABr and NaOTos. Relative to water the ortho protons shift slightly upfield (0.01 - 0.02 ppm); whereas the meta protons and those of the para methyl group shift appreciably upfield by ca. 0.17 ppm. The maximum shifts are observed at approximately equimolar CTABr and NaOTos (Figure 5). In addition the peaks broaden considerably; and with arene sulfonates, resolvable spectra are not obtained when the p-alkyl group is isopropyl or tert-butyl, even after 40 repetitive scans. In considering these shifts, it is perhaps best to focus attention on the relative proton shifts because of possible effects of the CTABr-tosylate mixtures on the chemical shift of the water protons; although the experiments of Eriksson and Gillberg on aqueous CTABr suggest that surfactant effects on the chemical shift of the water are relatively unimportant [33].

The NMR spectrum of CTABr changes when sodium arene sulfonates are added. The $(CH_3)_3N^+$ protons are broadened

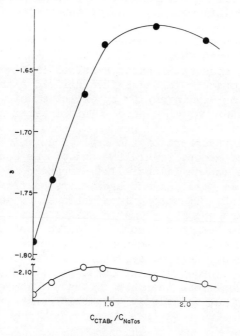

Figure 5. Downfield chemical shifts, relative to HOD, of the meta, ●, and ortho, ○, protons of 0.015 M NaOTos with added CTABr.

TABLE VI
CHEMICAL SHIFTS OF $\overset{+}{N}(CH_3)_3$ IN NaOTos-CTABr

C_{NaOTos}/C_{CTABr}	δ^a, ppm	$10^2 W_{1/2}$, ppm
	1.488	1.1
0.43	1.584	1.3
0.86	1.648	7.3
1.29	1.700	9.5
1.51	1.682	9.5
2.16	1.707	10.5

(a) Upfield shifts relative to D_2O/H_2O 90:10 in 0.0269 M CTABr at 31°.

---

and shifted upfield (Table VI), and the broadening increases with increasing hydrophobicity of the arene sulfonate. With added NaOTos, the methylene protons give two partially resolved peaks, one shifted upfield; the other downfield (Figure 6). With sodium benzene sulfonate, only one broad peak is observed; but neither salt has any marked effect on the terminal methyl group of the surfactant.

These data support the hypothesis that the phenyl group is inserted into the micelle so that the sulfonate group is in an aqueous environment as shown in Figure 3. The N-methyl protons should be shifted upfield either by the ring current of the aryl group or by exclusion of water molecules around the cationic head. This insertion also shifts some of the C-methylene proton signals upfield, because of the ring current of the aryl group. The protons of the C-methylene groups toward the end of the chain shift downfield because in the laminar-like structure, the long alkyl groups become parallel and are separated from each other. These relatively strong interactions between the cationic surfactant and the aromatic solute make the micellar structure larger and more rigid, so that the NMR signals are broadened. We examined the effects of some inorganic salts upon the chemical shift of the $(CH_3)_3\overset{+}{N}$ protons of CTABr. The effects of $NaNO_3$, NaOAc, NaCl, Na_4SO_4 and NaBr were less than 20% of that of NaOTos, and were observed only at relatively high salt concentrations (0.2 - 1.3 M).

Erickson and Gillberg pointed out some years ago that the association of aromatic solutes with cationic micelles was quite different from that of aliphatic solutes; for

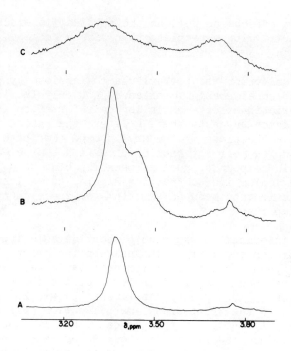

Figure 6. NMR spectra of the methylene and ω-methyl protons of 0.027 M CTABr, curve A; with 0.006 M NaOTos, curve B; with 0.029 M NaOTos, curve C.

example they concluded that a nonpolar hydrocarbon, e.g., isopropylbenzene, inserted into the micelle with the alkyl group toward the center; whereas a more polar aromatic compound, e.g., nitrobenzene, had the nitro group close to the micellar surface [33]. However in these systems, there was relatively little effect on the chemical shifts of the N-methyl protons, as would be expected if these nonionic solutes penetrated more deeply into the micelle than does an anion such as tosylate. The present results accord with this view. For example even relatively hydrophobic aliphatic carboxylate ions, such as pivalate, have no marked effects upon the NMR spectra of CTABr. Our experiments also confirm the importance of p-alkyl substituents; for example both viscosity and NMR experiments suggest that tosylate interacts much more strongly with micelles of CTA^+ than does benzene sulfonate ion, and the same is true of toluate as compared with benzoate ion. This mode of interaction appears to be general for these anions because kinetic salt effects as well as the physical interactions are more marked with a tosylate than with a benzene sulfonate ion.

This "solvation" of tetraalkylammonium ions by phenyl groups is of considerable importance, not only in micellar-solute interactions but also in ion pair formation, and our NMR evidence suggests that in water the cation is located above the plane of the aryl group. For both micellar and ion pair interactions, we find that electron rich aromatic compounds, such as phenols, bind to a tetraalkylammonium ion so that ion-dipole, as well as ion pair interactions, between an ammonium ion and a phenyl group are important.

These interactions are also of considerable importance in carbonium ion chemistry. The neighboring group participation of phenyl groups in S_N1 reactions is well established and has been explained in terms of the formation of phenonium ions [34]. Recent experiments by Mr. S. Huang on reactions of the tri-p-anisyl methyl cation (VI) with nucleophiles, X, show that phenols can stabilize the cation and reduce its reactivity towards nucleophiles, i.e. an intermolecular interaction between the tri-p-anisyl

$$(CH_3O-\langle\rangle)_3 C^+ + X \longrightarrow (CH_3O-\langle\rangle)_3 CX^+$$

VI

methyl cation and phenols is kinetically important in these reactions.

We are now in a position to try to explain the kinetic implications of these interactions between electrolyte and micelle. In the transition state for decarboxylation, the negative charge on the carboxylate group is dispersed into the cyclic π-system. We will assume that the planar benzisoxazole ring fits into the cationic micelle as shown in Figure 7, with the carboxylate group protruding into the water rich region around the micelle. With formation of the transition state, the negative charge will be dispersed into the cyclic system where it can interact strongly with the cationic head groups. Anions in the Stern layer of the micelle will have a beneficial effect on reaction rate because there will be electrostatic repulsions between them and the carboxylate group of the substrates which

ELECTROLYTE EFFECTS ON MICELLAR CATALYSIS

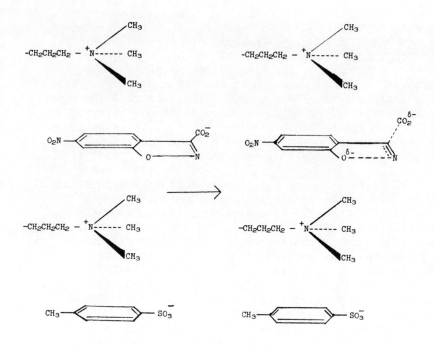

Figure 7. Model structures of the initial and transition states for the decarboxylation of 6-nitrobenzisoxazole-3-carboxylate ion in the CTA^+ $OTos^-$ comicelle.

will destabilize the initial state, and these unfavorable interactions will be relieved as the negative charge is dispersed during transition state formation.

An additional effect of tosylate ions is that with CTABr, they generate micelles with the rodlike structure which can readily accept the planar benzisoxazole carboxylate ion (Figure 7), and the unfavorable interactions between the tosylate and carboxylate ions will be relieved during formation of the transition state. However an excess of tosylate ions will generate an anionic micelle which will not take up the anionic substrate, so that the reaction rate then will fall sharply. On this hypothesis, we suppose that ions such as tosylate can affect the reaction rate in two ways; whereas smaller hydrophilic anions probably act simply by increasing the number of anions in the Stern layer of the cationic micelle and thereby destabilizing the initial state relative to the transition state.

The importance of the fit of the anion into the micelle is important; for example sodium β-naphthalene sulfonate, although more hydrophobic than sodium tosylate, has a smaller rate effect on the CTABr catalyzed decarboxylation (Table IV), possibly because it does not insert as easily into the cationic micelle; although it is better than tosylate as an inhibitor of the reaction of hydroxide ion with p-nitrophenyl diphenyl phosphate in CTABr [12].

This discussion of the kinetic role of tosylate ion has focused on the way in which the ion modifies the structure of a micelle by inserting between the cationic head groups and thereby changing the free energies of the initial and transition states of a decarboxylation. There is extensive evidence that nonionic aromatic solutes reside close to the surface of a cationic micelle [32,33], and therefore they should affect the catalyzed decarboxylation of 6-nitrobenzisoxazole-3-carboxylate ion. Benzene has almost no rate effect, but the behavior of phenols and phenoxide ions was unexpected because phenols reduce the reaction rate; whereas phenoxide ions increase it (Table VII). We assume that phenol and phenoxide ion will be incorporated into micelles of CTABr, and like the anionic solutes will perturb the micellar structure; although this point remains to be tested. Such a change of micellar structure could stabilize the initial state of the reaction, by allowing the benzisoxazole carboxylate ion to fit more readily into the micelle, and this initial state stabilization reduces reaction rate. Hydrogen

TABLE VII

EFFECT OF PHENOL AND PHENOXIDE ION ON THE CTABr CATALYZED DECARBOXYLATION OF 6-NITROBENZISOXAZOLE-3-CARBOXYLATE ION[a]

Solute	C_{solute}, M					
	0	0.01	0.02	0.03	0.033	0.06
$C_6H_5O^-$ [b]	3.71	4.25	4.09			3.39
$p\text{-MeC}_6H_4O^-$ [b]		4.03	4.03	4.59	4.84	3.86
C_6H_5OH [c]	3.38	2.47	1.85	1.60		0.95
$p\text{-MeC}_6H_4OH$ [c]		2.42		1.22		

(a) Values of $10^4 k_\psi$ sec^{-1}, at 25.0° in 2×10^{-2} M CTABr; (b) in 0.1 M NaOH; (c) at pH 7 0.002 M Tns.

bonding between the phenol and the benzisoxazole carboxylate ion in the micelle would further reduce the rate by stabilizing the initial state. Phenoxide ion should also modify the micellar structure, but the unfavorable electrostatic interactions between its charge and that of the carboxylate ion will be relieved in going from the initial to the transition state with an increase of reaction rate. Therefore with phenol and phenoxide ion, we have the paradoxical situation that the added anion assists cationic micellar catalysis, but its conjugate acid reduces it.

We have examined another example of salt effects upon a decarboxylation in the presence of micelles, but here we find the "normal" behavior. The decomposition of 3-bromo-3-phenyl propionate ion (VII) to olefin and lactone goes via a zwitterionic intermediate [35], and is retarded by micelles of CTABr.[29].

$$\underset{\text{VII}}{\text{PhCH(Br)} - \text{CH}_2\text{CO}_2^-} \xrightarrow{\text{slow}} \text{PhCH}^+ - \text{CH}_2\text{CO}_2^- + \text{Br}^-$$

$$\swarrow \qquad \searrow$$

$$\text{PhCH} = \text{CH}_2 \qquad \underset{\text{O} - \text{CO}}{\text{PhCH} - \text{CH}_2}$$

The reaction rate increases when salts are added to CTABr (Table VIII), simply because added anions tend to exclude the anionic substrate from the cationic micelle, so that more of it reacts in water, where the rate is greater than in the micelle. The more hydrophobic is the added anion,

TABLE VIII

SALT EFFECTS UPON THE CTABr INHIBITED REACTION OF 3-BROMO-3-PHENYL PROPIONATE ION[a]

C_{salt}, M	Salt			
	NaCl	NaOAc	Na_2SO_4	NaOTos
0.01				7.86
0.20	4.50	2.45	3.20	
0.60	5.01	3.30	3.30	

(a) Values of $10^2 k_\psi$, sec^{-1}. In 0.01 M CTABr $10^2 k_\psi$ = 1.66 sec^{-1}, in the absence of CTABr, $10^2 k_\psi$ = 14.7 sec^{-1} [29].

the more it excludes the substrate from the cationic micelle and the greater is the reaction rate. This reaction was followed by the changing ultraviolet absorption, and therefore we could not use salts with strong chromophores in this region.

There are a number of questions which we cannot answer as yet. (i) Why is decarboxylation of a benzisoxazole carboxylate ion, or a cyanoacetate ion, so different from other reactions catalyzed by cationic micelles, especially the spontaneous hydrolysis of dinitrophenyl phosphate dianions even though both reactions involve spontaneous, unimolecular dissociations? We can speculate that the difference depends on the extent to which negative charge is dispersed from a hydrophilic carboxylate or phosphate ion into a conjugated π-system in the transition state, and the fact that in phosphate ester hydrolysis, a dianion splits into a phenolate monoanion and a metaphosphate monoanion; whereas in decarboxylation, a carboxylate ion generates a planar, resonance stabilized, carbanion-like transition state. (ii) Why do some anions, e.g., cholate and thiocyanate, inhibit the micellar catalyzed decarboxylation at all concentrations?

There are now many examples of micellar catalysis and inhibition, but the work discussed here is a tentative attempt to control and understand these effects and especially the interactions between micelles and solutes.

This discussion has been concerned largely with the way in which electrolytes can perturb micellar structure and control micellar catalysis. However micellar structures are also modified by the incorporation of nonionic solutes, and the rates of micellar catalyzed reactions are often very sensitive to the addition of organic solvents. Often the concentration of the organic solvent, although low relative to that of water, is much larger than that of the surfactant; for example 1% by weight of dioxane gives a concentration of approximately 0.11 M, and this concentration has a marked effect on the hydrolysis of 2,4-dinitrophenyl phosphate dianion in CTABr [26]. But in other cases, relatively low molar concentrations of nonionic solutes can have appreciable kinetic effects; for example in the decarboxylation of 6-nitrobenzisoxazole-3-carboxylate ion in CTABr, testosterone and phenol although uncharged, decrease the rate as much as does cholate ion although the anion of estrone has almost no kinetic effect (Table IV).

We can expect to find many examples in which small amounts of ionic and nonionic solutes control micellar catalysis by modifying micellar structure, and we hope to be able to explain these effects in terms of structural changes, and thereby gain some understanding of the way in which relatively small molecules may control other catalyzed reactions, especially in biological systems.

I should like to acknowledge the contributions of Professors E. J. Fendler and Luis Sepulveda, Drs. L. Robinson, M. J. Minch and Messrs. Huang, Kamego and Kui-Un Yang, whose work I have discussed today; as well as those of my other colleagues who have worked with me on kinetic micellar effects. This work has been supported by the National Science Foundation and the Arthritis and Metabolic Diseases Institute of the U.S.P.H.S.

REFERENCES

(1) C. K. Ingold, "Structure and Mechanism in Organic Chemistry," 2nd Edn., Cornell University Press, Ithaca, N. Y., 1969.
(2) L. P. Hammett, "Physical Organic Chemistry," 2nd Edn., McGraw Hill, New York, 1970.
(3) R. W. Alder, R. Baker and J. M. Brown, "Mechanism in Organic Chemistry," Wiley-Interscience, New York, 1971.
(4) J. R. Cox and O. B. Ramsay, Chem. Rev., 64, 314 (1964).
(5) C. A. Bunton, Accounts of Chem. Res., 3, 257 (1970).
(6) A. J. Kirby and A. G. Varvoglis, J. Amer. Chem. Soc., 89, 413 (1967).
(7) C. A. Bunton, E. J. Fendler and J. H. Fendler, J. Amer. Chem. Soc., 89, 1221 (1967).
(8) C. A. Bunton, S. J. Farber and E. J. Fendler, J. Org. Chem., 33, 29 (1968).
(9) H. Morawetz, Advan. Catal. Relat. Subj., 20, 341 (1969); Accounts of Chem. Res., 3, 354 (1970).
(10) E. J. Fendler and J. H. Fendler, Advan. Phys. Org. Chem., 8, 271 (1970).
(11) E. H. Cordes and R. B. Dunlap. Accounts of Chem. Res., 2, 329 (1969).
(12) C. A. Bunton and L. Robinson, J. Org. Chem., 34, 773 (1969).
(13) S. T. Ahmad and S. Friberg, J. Amer. Chem. Soc., 94, 5196 (1972).
(14) T. C. Bruice and S. J. Benkovic, "Bioorganic Mechanisms," Benjamin, New York, 1966; W. P. Jencks, "Catalysis in Chemistry and Enzymology," McGraw-Hill, New York, 1969.
(15a) K. Shinoda, T. Nakagawa, B. Tamamushi and T. Isemura, "Colloidal Surfactants," Academic Press, New York, 1963; (b) P. H. Elworthy, A. T. Florence and C. B. Macfarlane, "Solubilization by Surface Active Agents and its Application in Chemistry and the Biological Sciences," Chapman Hall, London, 1968.
(16) H. Schott, J. Pharm. Sci., 60, 1594 (1971).
(17) R. B. Dunlap and E. H. Cordes, J. Amer. Chem. Soc., 90, 4395 (1968).
(18) C. A. Bunton and L. Robinson, J. Amer. Chem. Soc., 90, 5972 (1968); J. Org. Chem., 34, 780 (1969).
(19) J. L. Kurz, J. Phys. Chem., 66, 2239 (1962).
(20) L. R. Romsted and E. H. Cordes, J. Amer. Chem. Soc., 90, 4404 (1968).

(21) C. A. Bunton and L. Robinson, J. Amer. Chem. Soc., 91, 6072 (1969).
(22) C. A. Bunton and L. Robinson, J. Phys. Chem., 73, 4237 (1969); 74, 1062 (1970).
(23) E. W. Anacker and H. M. Ghose, J. Phys. Chem., 67, 1713 (1963); J. Amer. Chem. Soc., 90, 3161 (1968); T. Cohen and T. Vassilliades, J. Phys. Chem., 65, 1774 (1961).
(24) C. A. Bunton, E. J. Fendler, L. Sepulveda and K-U. Yang, J. Amer. Chem. Soc., 90, 5512 (1968).
(25) F. M. Menger and C. E. Portnoy, J. Amer. Chem. Soc., 89, 4968 (1967).
(26) G. J. Buist, C. A. Bunton, L. Robinson, L. Sepulveda and M. Stam, J. Amer. Chem. Soc., 92, 4072 (1970).
(27) C. A. Bunton and M. J. Minch, Tetrahedron Lett., 3881 (1970).
(28) C. A. Bunton, M. J. Minch and L. Sepulveda, J. Phys. Chem., 75, 2707 (1971).
(29) C. A. Bunton, A. Kamego and M. J. Minch, J. Org. Chem., 37, 1388 (1972).
(30) J. Baumbrucker, M. Calzadilla, M. Centeno, G. Lehrmann, P. Lindquist, D. Dunham, M. Price, E. Sears and E. H. Cordes, J. Phys. Chem., 74, 1152 (1970).
(31) Ref. 15b, Chapter 2.
(32) J. H. Fendler and L. K. Patterson, J. Phys. Chem., 74, 4608 (1970); 75, 3907 (1971).
(33) J. C. Eriksson and G. Gillberg, Acta Chem. Scand., 20, 2019 (1966).
(34) D. J. Cram, J. Amer. Chem. Soc., 74, 2129, 2137, 2159 (1952); C. J. Kim and H. C. Brown, J. Amer. Chem. Soc., 94, 5051 (1972) and ref. cited.
(35) F. G. Bordwell and A. C. Knipe, J. Org. Chem., 35, 2956, 2959 (1970).

MICELLAR CONTROL OF THE NITROUS ACID DEAMINATION REACTION

Robert A. Moss, Charles J. Talkowski,
David W. Reger, and Warren L. Sunshine
Wright Laboratory, School of Chemistry,
Rutgers University, The State University of
New Jersey, New Brunswick, N. J. 08903

The nitrous acid deamination of amines descends from Piria's discovery (1848) that the action of nitrous acid on either aspartic acid or asparagine gave malic acid.[1] Despite subsequent intensive investigation,[2] the important role which micelles can play in controlling the kinetics and stereochemistry of this reaction remained unrevealed until 1969.[3]

We now believe that the micelles perturb, rather than usurp, the "normal" mechanism of the deamination reaction, and we therefore begin with a discussion of that mechanism. A rather didactic interpretation appears in Scheme I.[2e]

Scheme I

$$2HNO_2 \underset{}{\overset{fast}{\rightleftharpoons}} N_2O_3 + H_2O \quad (1)$$

$$RNH_2 + N_2O_3 \xrightarrow{slow} R\overset{+}{N}H_2NO + NO_2^- \quad (2)$$

$$R\overset{+}{N}H_2NO \xrightarrow[H_2O]{fast} RN=N-OH + H_3O^+ \quad (3)$$

$$\left[\underset{I}{R-N=N-OH} \underset{-H^+}{\overset{+H^+}{\rightleftharpoons}} \underset{II}{R-N=N-\overset{+}{O}H_2} \underset{+H_2O}{\overset{-H_2O}{\rightleftharpoons}} \underset{III}{RN_2^+} \rightleftharpoons \underset{IV}{R^+} + N_2 \right] \quad (4)$$

$$\downarrow H_2O$$

Alcohols, Alkenes

The kinetic behavior reflects the rate determining nitrosation step (2), and is third order because of the pre-equilibrium (1) which generates the actual nitrosating agent, nitrous anhydride.[2b,4-6] Note that nothing which we might learn about the reaction kinetics can tell us anything about the microscopic mechanism of product formation.

The "kinetic phase" of the sequence (2) is connected, <u>via</u> rapid proton transfers and tautomerism (3) with the product determining phase (4). The latter is initiated with the diazotic acid, I, which is linked by protolysis to the diazonium ion, III, and thence, by loss of nitrogen, to the carbonium ion, IV.

Duels have been fought over which one (or several) of the species I-IV is responsible for product formation.[2] Though we will consider this problem below, we note now that we hope to study the blend of discrete molecular acts combined in (4) by following the <u>stereochemistry</u> of the amine to alcohol transformation.

KINETICS

We began by deaminating 2-aminoalkanes in 10% (1.6<u>M</u>) aqueous sodium nitrite solution, at 25° and pH 4. The pH was adjusted and maintained by the addition of strong acid, initially perchloric acid. These constitute <u>standard conditions</u>.

The reaction rate was monitored manometrically, after the evolved nitrogen had been scrubbed clean of nitrogen oxides and low-boiling alkenes.[7] The deamination reactions obeyed rate law (5).

$$d[N_2]/dt = k_3[RNH_2][NO_2^-]^2 \qquad (5)$$

For 2-aminobutane through 2-aminohexane, (5) held for 20-40% of the reaction, over a range of amine and nitrite concentrations. Eventually, k_3 deviated negatively, presumably because of non-deaminative destruction of nitrous acid.[4]

Due to the excess of nitrite ion present, the deaminations were also pseudo first order in amine. For example, 2-aminobutane at five initial concentrations (0.16-0.68<u>M</u>) gave k_1=0.044 ± 0.003 min^{-1} and k_3=2.2x10^{-2} $\ell.^2$mol^{-2}min^{-1}. The calculated k_1 (from k_3 and [NO$_2^-$]^2) was 0.048 min^{-1}.

For convenience, we will discuss the kinetics in terms of k_1 where

$$d[N_2]/dt = k_1[RNH_2] \qquad (6)$$

$$k_1 \sim k_3[NO_2^-]^2 \qquad (7)$$

The relation of k_1 and k_3, (7), was satisfied at many initial nitrite concentrations. We thus obtained the data shown in Table 1.

Table 1
Kinetic Results: Deamination of $RCH(NH_2)CH_3$ [a]

R	$[RNH_2]$, $\underline{M}$	k_1, min^{-1}	Cmc, $\underline{M}$ [b]
C_2H_5	0.16 – 0.68	$0.043 \pm 0.003_6$	------
$\underline{n}$-C_3H_7	0.23 – 0.68	$0.033 \pm 0.002_6$	> 1.81
$\underline{n}$-C_4H_9	0.20 – 0.64	$0.031 \pm 0.001_3$	0.89
$\underline{n}$-C_5H_{11}	0.20	$0.034 \pm 0.002_3$	0.24

[a]Standard conditions, see text. [b]Critical micelle concentrations of $RCH(NH_3^+)CH_3$ in water, $1.6\underline{M}$ NaClO$_4$, pH4, HClO$_4$, determined tensiometrically.

Note that as long as the initial alkylammonium ion[8] concentration is <u>below</u> the cmc, the observed rate constants are essentially independent of initial concentration and chain length. But, as demonstrated in Table 2, these "constants" are augmented whenever the amines are deaminated at concentrations which exceed the systemic cmc.[9]

Table 2

Kinetic Results: Deamination of $RCH(NH_2)CH_3$ Above the Cmc[a]

R	[RNH_2], $\underline{M}$	k_1, min^{-1}	Cmc, $\underline{M}$
$\underline{n}$-C_5H_{11}	0.36	0.22 ± 0.01$_3$	0.24
$\underline{n}$-C_6H_{13}	0.23	0.37	0.058
$\underline{n}$-C_6H_{13}	0.24	0.46	0.058
$\underline{n}$-C_6H_{13}	0.44	0.55	0.058

[a]Conditions as in Table 1.

The observed k_1 is greater by a concentration dependent factor of ~6-15, although relations (5) - (7) continue to hold. If we consider k_{obs} to be the sum of micellar (k_m) and "free" (k_f) deaminative rate constants, then we can estimate that k_m for the 2-aminooctane system is 0.5-0.6 min^{-1} (see below).

Moreover, micellar catalysis of deamination can be induced in mixed micelles composed of the ammonium ion to be deaminated and inert alkyltrimethylammonium ions, $\underline{cf}$. Table 3.

Note that the 2-aminoheptane deaminates at a normal rate when its concentration is below the systemic cmc (Case 1). Addition of the simple salt, NaBr, causes no change (Case 2), and the addition of a small ammonium ion (Case 3) has only a marginal effect. However, the addition of the longer chain, and micellized,[10] OTA$^+$ Br$^-$ (Case 5) induces a marked enhancement of the observed k_1. The twofold acceleration noted in Case 4 presumably indicates that the total initial alkylammonium ion concentration exceeds the systemic cmc by only a small factor.

Table 3

Catalysis of 2-Aminoheptane Deamination by Permanent Micelles[a]

Case	[RNH$_2$], M	Additive, M	k_1, min^{-1}
1	0.20	none	0.034
2	0.18	NaBr, 0.29	0.030
3	0.18	Et$_4$N$^+$Br$^-$, 0.27	0.043
4	0.18	OTA$^+$Br$^-$, 0.11[b]	0.075[c]
5	0.18	OTA$^+$Br$^-$, 0.28	0.26[c,d]

[a]Standard conditions. [b]OTA$^+$Br$^-$ is 2-octyltrimethylammonium bromide. [c]HCl replaced HClO$_4$. Both acids gave comparable results in Case 1, but OTA$^+$Br$^-$ gives the insoluble OTA$^+$ClO$_4^-$ with HClO$_4$. [d]Unaffected by the addition of 0.29M NaBr.

Not surprisingly, this type of micellar catalysis is rather ineffective with short chain amines, which should have low affinities for the micellar phases. Under the conditions of Case 5, for example, k_1 of 2-aminobutane is only 0.087, compared to the non-micellar value of 0.043 min^{-1} (Table 1).

What is the origin of the micellar catalysis? The rate determining nitrosation (2) generates a positive charge on the micelle-solubilized substrate, and is thus unlikely to be made more facile by a cationic micelle. Therefore the favorable electrostatic stabilization of an oppositely charged transition state by an ionic micelle, a factor often operative in micellar catalysis,[12] plays no role in the nitrosation step.

Consider, however, the stoichiometric rate law (5). In reality, we know that N_2O_3 is the nitrosating agent, and that free amine, not its ammonium ion, is the substrate. We can therefore write (8):

$$d[N_2]/dt = k_{real}[N_2O_3][RNH_2] \qquad (8)$$

Equation (8) can be converted to (9) by insertion of the proper equilibrium constants and concentration terms.

$$\frac{d[N_2]}{dt} = \frac{k_{real} K_{eq}^{N_2O_3} [NO_2^-]^2 [H^+] K_A^{RNH_2} [RNH_3^+]}{(K_A^{HNO_2})^2} \qquad (9)$$

Here, k_{real} is the rate constant for the reaction of nitrous anhydride with the free amine; $K_{eq}^{N_2O_3}$ is the equilibrium constant for the formation of the anhydride from nitrous acid; and the acidity constants have their usual meanings. From (5) and (9) we derive (10).[13]

$$k_3 = \frac{k_{real} K_{eq}^{N_2O_3} K_A^{RNH_3^+} [H^+]}{(K_A^{HNO_2})^2} \qquad (10)$$

The observed rate constant, k_3, is dependent on several equilibrium constants which precede the rate step. How are these "constants" affected by micellization? The positive-positive repulsions within the Stern layer of the micellar ammonium ions will destabilize protonated amine relative to the free base, and $K_A^{RNH_3^+}$ will surely be augmented on micellization of the ammonium ions. Indeed, Cordes has reported just such an effect attending the solubilization of protonated iminium ions in cetyltrimethylammonium bromide micelles.[14] K_A was enhanced by a factor of ~15, in this instance, sufficient to account for our observed catalytic effect.

Thus, one explanation for the catalysis is simply that, at constant pH, micellization of the alkylammonium ions increases the available free amine,[15] nitrosation is more rapid, and the enhancement appears in a larger measured k_3.

To be thorough, we must examine the response to micellization of each term of (10). Electrostatic micellar effects on $K_A^{HNO_2}$ should be minimal, because the equilibrium is of the neutral ⇌ ion pair type. Similarly, we expect little perturbation of the nitrous acid-nitrous anhydride equilibrium. However, we have no special knowledge concerning the possibility of micellar solubilization of nitrous anhydride.

We presently believe that micellar enhancement of alkylammonium ion acidity is the major determinant of the micellar catalysis of the nitrous acid deamination reaction. In accord with this concept, we recall that other cationic micelles can elicit the catalysis. Moreover, we might not anticipate a very pronounced dependence of the catalytic effect on the nature of the associated counterions (anions), because the catalysis finally depends on a neutral+neutral reaction within a cationic micelle.

STEREOCHEMISTRY

The product forming phase of the deamination reaction suffers from a plethora of possible precursors. Because this is not the place to consider in detail the claims favoring one or another of species I-IV, we will, at the cost of some balance, posit that the alkyldiazonium ion stage (III) is not attained from sec-alkyldiazotic acids (I) in acidic aqueous solution. That is, C-N bond breaking occurs in reasonable concert with N-O bond breaking in the decomposition of secondary I or II.[16]

The product alcohol therefore arises by (a) the attack of water on I or II; (b) the attack of water on V, an alkyl cation in an ocean of water, separated from the original diazotic acid oxygen atom by the departing nitrogen molecule; (c) the frontside return of the original water molecule within V (or frontside attack of a water molecule which is hydrogen-bonded to it); or (d) the drowning of a symmetrically "hydrated" alkyl cation derived from V. These

$$[R^+{}^{N\equiv N}OH_2]_{H_2O}$$

V

possibilities are outlined in Scheme II, which implies that the weighted sum of the competing pathways will generate the overall stereochemical result of any particular amine→alcohol deaminative transformation.

Scheme II[a]

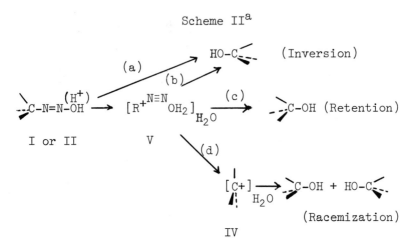

[a]The letters designating the pathways refer to the text.

    An analogous scheme could, of course, be drawn for any solvolytic process, e.g., the hydrolysis of 2-octyl mesylate. But in such a process, the activation energy for the C-O fission step would be ~25-30 Kcal/mole, and the differential activation energies between the pathways of the scheme would probably be large, ca. 2 Kcal/mole. Effective competition would not occur, and a pure stereochemical result would be obtained. Indeed, the hydrolysis of 2-octyl mesylate does take place with 99% inversion.[17]

    Conversely, the product forming stage of the deamination reaction has a low activation energy, estimated at 3-5 Kcal/mole.[2a] Now the differential activation energies in Scheme II will also be low (<1 Kcal/mole), and effective competition should be possible. We can therefore expect strong dependence of the reaction stereochemistry on small alterations in alkyl group structure and solvent polarity, for these will affect the ease of the I (or II) to V conversion. (We imply that stereochemical alternatives to inversion must begin with V.)

    How well does Scheme II rationalize the literature? The primary carbinamine, 1-aminobutane-1-d, is converted to the corresponding acetate with ~69% net inversion in nitrous acid-acetic acid,[18] suggesting much direct attack of acetic acid on the butane-1-diazotic acid (or n-butyldiazonium ion) derived from it. This is reasonable, for primary carbinamine derivatives would be relatively slow to give species such as V, in which positive charge resides on a primary carbon atom. Most of the 1-butyl acetate should indeed come from inverting displacement on the diazotic acid (Scheme II, path a).

Table 4 gathers contrasting data reported for sec-carbinamine deaminations. Note that 2-aminobutane affords 2-butanol with considerably less net inversion (22%) than was observed in the 1-aminobutane reaction (69%).[18] The butane-2-diazotic acid can afford a secondary carbonium ion, rather than a primary carbonium ion. Now the I→ V conversion is faster, and more competitive with inverting solvolysis of I.

Table 4

Stereochemistry of Deamination of sec-Carbinamines[a]

R in $RCH(NH_2)CH_3$	% Net Inversion in $RCH(OH)CH_3$
C_2H_5	22[b]
C_6H_5	14[c]
c-C_3H_5	0-5[d]
n-C_6H_{13}	0-2[e]

[a]Aqueous nitrous acid, pH 4. [b]References 19 and 20. [c]Reference 20. [d]Reference 21. [e]At ~0.4M amine, reference 20.

Moreover, paths c and d will be competitive with b. In sum, less inversion is both expected and observed in the deamination of a sec- compared with a prim- carbinamine.

Extensions of these arguments rationalize the even greater approach to racemization demonstrated in the deaminations of α-phenylethylamine and α-cyclopropylethylamine (Table 4): the order of decreasing net inversion parallels the order of increasing potential carbonium ion stability.

However, we were completely unprepared for the observation of <u>racemization</u> attending the deamination of 2-aminooctane.[20] Nothing in Scheme II or in carbonium ion theory implied that the 2-octyl system should exhibit behavior markedly different from that of the 2-butyl system. To add to the confusion, the literature maintained that the 2-aminooctane→2-octanol conversion occurred with "inversion + racemization."[22] We know now that this "anomalous" stereochemistry reflects micellar phenomena,[3] which can be clearly revealed upon examination of an homologous series of optically active 2-aminoalkanes.

The stereochemical outcomes of deamination reactions within such a series are gathered in Table 5.

Table 5

Stereochemistry of 2-Alkanol Formation in the Deamination of $RCH(NH_2)CH_3$ [a]

R	Stereochemistry ± av. dev. (% net)		n	Cmc,$\underline{M}$ [b]
C_2H_5 [c]	22.9	inv.	0.4_3(%)	-----
$\underline{n}$-C_3H_7	22.9	inv.	2.2_3	>1.81
$\underline{n}$-C_4H_9	19.0	inv.	0.6_2	0.89[d]
$\underline{n}$-C_5H_{11}	3.5	inv.	0.4_4	0.24
$\underline{n}$-C_6H_{13}	6.0	ret.	e	0.058
$\underline{n}$-C_8H_{17}	11.8	ret.	0.5_2	-----

[a][RNH_2] = 0.76$\underline{M}$; standard conditions. [b]Cmc's of racemic $RCH(NH_3^+)CH_3$, determined in 1.6$\underline{M}$ $NaClO_4$ solution, at pH 4 ($HClO_4$), by surface tensiometry. The av. dev. of these determinations was ±0.02$\underline{M}$. [c]Initial amine concentrations were 0.4 or 0.6$\underline{M}$. [d]Cmc with 1.6$\underline{M}$ $NaNO_3$, 0.83; NaCl, 0.87$\underline{M}$. [e]See below.

The amines were resolved via their tartrate salts,[23] and optical purities were determined by gc analyses of derived diastereomeric trifluoroacetyl-L-prolylamides.[24] Gc was used to isolate the optically stable product alcohols, the optical purities of which were then polarimetrically determined.[25] In the case of 2-octanol, we employed either polarimetry or gc analysis of the derived 2-octyl acetyllactates.[26]

Note, in Table 5, the dependence of stereochemistry on chain length initiated at $R>C_4H_9$. The stereochemistry deviated strongly from the norm (~23% net inv) and moved toward retention as soon as the iniaial ammonium ion concentration exceeded the cmc.[3] At $[RNH_2] = 0.76\underline{M}$, 2-aminohexane was the borderline case; had we worked at a constant initial concentration near $0.2\underline{M}$, 2-aminoheptane would have been the borderline case.

The "chain-length effect" results simply from the steady decrease in the cmc's of alkylammonium ions with increasing R. As R is made longer, more of the total deamination reaction (at a given initial amine concentration, chosen such that $C_i>cmc$) occurs in the micellar phase.

Although we have implicated micelles as agents of stereochemical control, we have not yet indicated the mode of their action. This will be discussed shortly, but we can note here that the stereochemical course of a homogeneous deamination reaction is known to move toward retention, with a decrease in solvent polarity, due to greater participation of return pathways (cf., Scheme II, c).[2c-e] Because the micellar phase, even at the Stern Layer, is less polar than water,[27] the outline of a possible mechanism for micellar stereochemical control is apparent.

A more useful way of demonstrating the micellar control involves studying the stereochemical dependence as a function of the initial concentration of a single amine. In contrast to the demonstration summarized by Table 5, the new method directly adjusts the fraction of the reaction which occurs in the presence of alkylammonium ion micelles, because the cmc is now a "constant".

Application of this method to the deamination of 2-aminooctane yields the results shown in Table 6. We see that as long as the initial 2-octylammonium ion concentration remains below the systemic cmc, the stereochemistry is normal, and similar to that observed for the short chain amines of Table 5.

If $[RNH_3^+]_i$ > cmc, however, the stereochemistry moves toward retention. At $[RNH_3^+]_i$ = 0.39$\underline{M}$, the concentration we chanced to select in our initial study[20] (cf., Table 4), the stereochemical result happens to be racemization. It was the fortunate choice of this concentration, which led to a discrepancy with Ingold's earlier work,[22] and, in turn, to all of the work presented here! At still higher initial concentrations, net retention is the outcome, and the 6% achieved at C_i = 0.76$\underline{M}$ is probably close to the limiting micellar stereochemistry for the deamination of 2-aminooctane.

Table 6

Deamination of 2-Aminooctane: The Concentration Dependence of Stereochemistry[a]

$[RNH_2]_i$, $\underline{M}$	Stereochemistry of $RNH_2 \rightarrow ROH$	
0.76	6. (% net)	ret.
0.39	0.	rac.
0.15	7.	inv.
0.083	16.3	inv.
0.076	18.	inv.
cmc[b]	-------------------	
0.029	25.	inv.
0.015	23.	inv.

[a]Standard conditions were employed. The 2-octanol optical purities were determined by gc, and each entry is the (averaged) result of at least two experiments. [b]The cmc is taken as 0.058$\underline{M}$; see below.

Scheme III presents a dynamic picture of the deamination reaction. The amine is in equilibrium with the corresponding ammonium ion, and it is more than 99% protonated at pH 4. If the ammonium ion concentration exceeds the cmc, then the excess goes into the micellar phase. The deaminatable form, the free amine, is found mainly in this phase (see above).

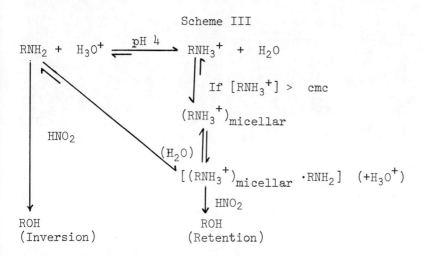

The alkylamine is deaminated both within the micellar phase and within the aqueous phase. For 2-aminooctane, we assign stereochemistries of ~6% net retention and ~24% net inversion for 2-octanol formed via micellar and non-micellar deaminations, respectively (see Table 6). The observable, <u>overall</u> stereochemistry will then be a weighted sum of the stereochemistries for the micellar and non-micellar deaminations, assuming that the two processes are comparably efficient[28] in the amine→ alcohol transformation.

Moreover, from Tables 1 and 2, we know that micellar deamination is faster than its non-micellar competitor, and we can tentatively adopt a model in which the overall deamination occurs in stages: the micellar deamination (6% net retention) proceeds until the ammonium ion titer falls below the cmc, then the non-micellar deamination (24% net inversion) completes the destruction of the substrate.

This model can be used to anticipate the stereochemistry of a specific deamination. We need to know the initial amine concentration, C_i, and the systemic cmc. Define a fraction of the reaction, F_m, which is micellar:

$$F_m \sim (C_i - cmc)/C_i \qquad (11)$$

and a fraction of the reaction, F_{non}, which is non-micellar.

$$F_{non} \sim cmc/C_i = 1-F_m \qquad (12)$$

Taking cmc = 0.058M for 2-octylammonium ion (under standard conditions), we can evaluate F_m and F_{non} for each C_i of Table 6. We then calculate the stereochemical contributions, at that C_i, of the micellar and non-micellar processes:

$$\text{Micellar Contribution} = F_m(53R + 47I) \qquad (13)$$
$$[6\% \text{ net ret.}]$$

$$\text{Non-micellar Contribution} = (1-F_m)(38R + 62I) \qquad (14)$$
$$[24\% \text{ net inv.}]$$

The R's and I's are summed, and the difference is taken, affording a predicted stereochemistry for C_i values of Table 6.[29] Predicted and observed results are compared in Table 7.

Table 7

Deamination of 2-Aminooctane: Predicted and Observed Stereochemistry[a]

C_i, M	F_m[b]	Stereochemistry[c] Observed[d]		Calculated	
0.39	0.85	0.0[e]		1.4	ret
0.15	0.61	7.0	inv	5.6	inv
0.083	0.30	16.3	inv	15.	inv
0.076	0.24	18.	inv	17.	inv

[a]Standard conditions. [b]See (11). [c]% net. [d]From Table 6. [e]Racemization.

Considering the model's simplicity, the agreement between calculated and observed stereochemistries is good. Note that the observed values are always a bit to the inversion side of the calculated values. This is understandable, because the model neglects the small contribution from the acqueous phase deamination which occurs <u>simultaneously</u> with the micellar deamination. The former process affords 2-octanol with 24% net inversion, and any incursion <u>during</u> the micellar stage will tip the actual result somewhat more toward inversion than we can calculate with our model.

We now have an overview of micellar control in the nitrous acid deamination reaction, including a semi-quantitative model of what is clearly a complex blend of many discrete and differing molecular acts. But we lack any microscopic, and therefore truly mechanistic, picture of the reaction's product forming stage; the stage which determines stereochemistry. A further advance can be made by investigating the influence of counterion identity on the micellar phenomena.

THE ANION DEPENDENCE OF MICELLAR CONTROL

1. Critical Micelle Concentrations

C_i, the initial alkylammonium ion concentration, and the cmc under the reaction conditions are the basic and required input for our micellar control model. Because the cmc of cationic micelles depends strongly on the concentration <u>and</u> the identity of anions present,[30] we need to be able to measure the cmc of the substrate ammonium ions under the deamination conditions. This is not possible, for, under the actual reaction conditions, these ions are rapidly being converted to nitrogen and products. A model reactant system is required in order to determine the needed cmc values.

In constructing the model, the principal problem is finding a satisfactory anion to stand in for the 1.6<u>M</u> nitrite which is normally present in the deamination.[31] 2-Octyltrimethylammonium (OTA) salts, which do not deaminate in acidic nitrite solution, were therefore synthesized. We prepared $OTA^+NO_2^-$, OTA^+Cl^-, OTA^+Br^-, and $OTA^+NO_3^-$, and then determined the cmc's of each of these salts as a function of total anion titer (supplied as NaX). Graphs of cmc vs. $[X^-]$ indicated that the behavior of $OTA^+NO_2^-$ was most closely mimicked by that of OTA^+Cl^-. This is not surprising, for chloride and nitrite are neighbors in the lyotropic series,[32] which has important applicability to such micellar phenomena as cmc depression.[30]

We then made simulated reactant systems which contained 1.6$\underline{M}$ NaCl (representing $NaNO_2$), alkylamine at a chosen C_i (e.$\underline{g}$., 0.15$\underline{M}$), and HX, added until pH 4 was obtained (~0.15$\underline{M}$, in this example). The cmc's of these systems were measured by continually observing their surface tensions as they were diluted with aqueous titrants containing 1.6$\underline{M}$ NaCl, 0.15$\underline{M}$ NaX, and HX (to pH 4). Under actual deamination conditions, HX neutralizes some of the nitrite present, as well as the amine, introducing "extra" X^-. In some of our model cmc studies, therefore, we set $[NaX] > [RNH_3^+]$ to account for this.

Cmc's determined in this manner are subject to some variation as a function of the particular C_i chosen for the experiment, for this choice also determines $[X^-]$. But Table 8 demonstrates that the dependence of cmc on C_i is not large, at least in the concentration ranges of interest to us.

Table 8

Cmc's Under "Deamination" Conditions: 2-Octylammonium Ion in 1.6M Aqueous NaCl.[a]

$[RNH_3^+]_i$,$\underline{M}$	$[HX]$,$\underline{M}$[b]	Titrant,$\underline{M}$	Cmc
0.60	$HClO_4$, 0.60	$NaClO_4$, 0.60 NaCl, 1.6	0.054
0.15	$HClO_4$, 0.15	$NaClO_4$, 0.15 NaCl, 1.6	0.061
0.80	HCl, 0.80	NaCl, 2.4	0.082
0.20	HCl, 0.20	NaCl, 1.8	0.078

[a] Only several examples, chosen from numerous runs, are listed. [b] At pH 4.

We next determined the cmc of 2-octylammonium ion in the presence of various other counterions, under "deamination conditions". The data are recorded in Table 9, in which the values for perchlorate and chloride are "best values", the result of many experiments at various amine and anion concentrations. The chloride value includes an additional correction to account for the difference in cmc-lowering power of nitrite ion and chloride ion. The cmc's of the other anions were determined at only one initial amine concentration; they should be regarded as approximate rather than definitive.

Table 9

Cmc's of 2-Octylammonium Ions Under Deamination Conditions with Added Anions.[a]

Anion	Concentration, $\underline{M}$	Cmc, $\underline{M}$
Cl^-	b	0.090
CH_3COO^-	0.59	0.089
BF_4^-	0.71	0.075
Br^-	0.53	0.073
ClO_4^-	b	0.058
$\underline{d}$-10-Camphorsulfonate	0.53	0.043
$\underline{p}$-Toluenesulfonate[c]	0.43	0.029

[a]In 1.6$\underline{M}$ aqueous NaCl solution, at 25° pH adjusted to 4.0 with HAnion. [b]"Best" value, see text. [c]At 35°.

With the cmc values of Table 9, we can apply our model for the micellar control of deamination to the study of anion dependence. We thus hope to dissect stereochemical (and kinetic) effects due simply to cmc variation as a function of anion selection, from more interesting effects which implicate cooperative anion-micelle control of the reaction mechanism.

2. Stereochemistry and Kinetics

We know that the kinetics (k_{obs}) and stereochemistry of the 2-aminooctane → 2-octanol transformation depend on $[RNH_2]_i$ in a fashion that implicates micellar control. More exactly, each reaction parameter gives a linear correlation with F_m (11). From the stereochemical data of Tables 6 and 7, and from the kinetic data of Tables 2 and 10, correlations 1 and 3 of Figure 1 were drawn.

Table 10

Kinetic Results: Deamination of 2-Aminooctane[a]

C_i, M	F_m[b]	k_{obs}(min^{-1})[c]
0.030	d	0.037
0.110	0.472	0.258
0.127	0.543	0.353
0.220	0.736	0.410
0.240	0.757	0.460
0.439	0.868	0.502

[a]Standard conditions ($HClO_4$). [b]See (11). [c]Pseudo first order rate constants are given. [d]Below the cmc (0.058 M).

The adequate correlations for these sodium nitrite-perchloric acid deaminations are apparent, though the stereochemistry should be only crudely linear in F_m. (See above for a fuller discussion.)

Analogous studies were next made with HCl (rather than $HClO_4$) as the acid source for the 2-aminooctane deamination.[33] With the appropriate cmc from Table 9, we calculated F_m values and drew correlations 2 and 4 of Figure 1.

At once, the dramatic result is perceived. The deamination reactions are accelerated by the 2-octylammonium micelles whatever the counterion, but the stereochemistry of

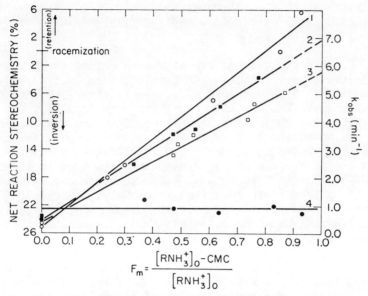

Figure 1. The stereochemistry and kinetics of the nitrous acid deamination of 2-aminooctane: curve 1, stereochemistry, perchlorate counterions; curve 2, kinetics, chloride counterions; curve 3, kinetics, perchlorate counterions; curve 4, stereochemistry, chloride counterions. The rate constants, as shown on the right-hand ordinate, have been arbitrarily multiplied by 10.0.

2-octanol formation is only altered if both micelles and perchlorate ions are present. Specific anions are necessary for the micellar control of stereochemistry.[34]

Various experiments were done to highlight this need. The (perchlorate) deamination of 2-aminooctane at C_i=0.083M (F_m=0.30) gave 2-octanol with 16.3% net inversion. However, in the presence of d,ℓ-2-aminodecane (0.80M, F_m>0.94), the deamination of 2-aminooctane at C_i=0.086M gave 2-octanol with 5.4% net retention. This stereochemistry is characteristic of a 2-aminooctane deamination at C_i=0.76M and F_m>0.9. We conclude that the 2-decylammonium micelles provided a "template" for 2-aminooctane deamination, which therefore occurred with a stereochemistry characteristic of F_m>0.9. However, perchlorate ion was essential; parallel experiments with chloride counterions failed to elicit a template effect, and 21-22% ("normal") net inversion was observed under all conditions.

A perchlorate deamination of 2-aminooctane (C_i=0.089$\underline{M}$, F_m=0.35) which was flooded with 2-aminobutane (0.89$\underline{M}$) gave 2-octanol with 9.3% net inversion. We do not believe that 2-butylammonium ion micellar templates were involved, but rather that the cmc of the 2-octylammonium ions was depressed by the added short-chain ammonium ions. Indeed, the cmc of 2-octylammonium ions, under our model reaction conditions and in the presence of 0.89$\underline{M}$ 2-butylammonium ion, is lowered from 0.058 (Table 9) to 0.0$\overline{48}\underline{M}$. With this new cmc, and the calculation proceedure described previously, we predict 10% net inversion as the stereochemical outcome of the 2-aminobutane flooding experiment, in excellent agreement with the observed result.

Inclusion of perchlorate ion in nitrite-hydrochloric acid deaminations of 2-aminooctane (F_m>0) shifted the reaction stereochemistry toward increased retention; added perchlorate did not further affect the stereochemistry of nitrite-perchloric acid deaminations. Apparently, the micellar control is already "saturated" at $[RNH_3^+]/[ClO_4^-]$=1.

The contrasting behaviors of chloride and perchlorate led us to examine other anions, and to classify each one by its ability to assist in micellar control of the 2-aminooctane deamination. The results (Table 11) show that bromide[35] and acetate join chloride as anions incapable of eliciting micellar stereochemical control. On the other hand, fluoroborate, tosylate, and 10-camphorsulfonate resemble perchlorate and do elicit control. The potency order appears to be (15).

$$ClO_4^- > \underline{p}\text{-Tos}^- > BF_4^- > CamSO_3^- \qquad (15)$$

Note especially that deamination (F_m>cmc) in the presence of $\underline{any}$ of the anions was accelerated, whether or not stereochemical control was manifested. From Table 11 we see that the potency order for kinetic control (16) appears to differ from (15).

Table 11

Micellar Deamination of 2-Aminooctane:
Anion Dependence of Kinetics and Stereochemistry[a]

Anion	$[RNH_2]_i$, $\underline{M}$	F_m[b]	Stereochemistry (% net inv.)		k_{obs}(min^{-1})	
			observed	normal[c]	observed	normal[c]
Br$^-$	0.340	0.79	22.2	23.4	0.420	0.031
CH$_3$COO$^-$	0.300	0.70	23.4	23.9	0.400	0.035
$\underline{d}$-C$_{10}$H$_{16}$SO$_4^-$[d]	0.380	0.89	17.3	25.4	0.252	0.044
BF$_4^-$	0.410	0.82	14.7	23.1	0.200[e]	0.036
$\underline{p}$-C$_7$H$_7$SO$_3^-$[f]	0.360	0.92	7.3	23.3	0.133[g]	0.030

[a]Standard deamination conditions were used; pH 4 was adjusted with HAnion. [b]Calculated from (11), using the indicated initial amine concentration and the appropriate cmc value from Table 9. [c]C_i < cmc. [d]$\underline{d}$-10-Camphorsulfonate. [e]At C_i=0.422$\underline{M}$, F=0.82. [f]$\underline{p}$-Toluenesulfonate. This reaction was done at 29°. [g]At C_i=0.306$\underline{M}$, F_m=0.90.

$$Cl^- \sim ClO_4^- \sim Br^- \sim OAc^- > CamSO_3^- \sim BF_4^- > \underline{p}\text{-Tos}^- \qquad (16)$$

We concluded above that the micellar kinetic control is electrostatic in origin and depends mainly on the cationic nature of the micelles. A minor dependence on counterion identity was anticipated. The kinetic results (Table 11) are generally in keeping with expectation, though the dependence is perhaps a bit larger than "minor", and the potency order is not yet rationalized. Nevertheless, all of the micellar deaminations are faster than their nonmicellar counterparts.

But how are we to understand the anion-stereochemical dependence of micellar deamination, in which the counterions appear to be separable into effective and ineffective classes?[36] In our discussion of Scheme II, we contended that the stereochemistry of alkanol formation was determined by the partition of R-N=N-OH between nitrogen loss with return of OH, or its equivalent (retention); displacement by water (inversion); and escape to hydrated alkyl cations (racemization). The partition's resultant is ~24% net inversion in aqueous solution, if R=2-alkyl.

We suggest that an identical partition occurs in the highly aqueous Stern layers of those alkylammonium ion micelles which are rather <u>weakly associated</u> with their counterions,[37] and that the stereochemistry of deamination reactions occurring within such micelles will not differ appreciably from comparable reactions in the bulk aqueous phase.[38]

From 50-70% of the counterions are normally within the micelle's Stern layer,[12,39] but a trend toward the upper limit is expressed as the hydrophobicity of the counterions is increased (<u>i.e.</u>, as their hydration energy is decreased). This view is supported by studies of light scattering by micelles,[30,37] and by studies of comparative counterion inhibition of micelle catalyzed ester hydrolyses.[12]

Moreover, poorly hydrated[40] counterions will be strongly bound, and will engender larger, more effectively charge-neutralized, and (probably) denser, less aqueous micelles.[37] It is to be expected that in such an altered environment, the partition which determines deamination stereochemistry will be biased toward return of OH and increased retention.[2d,e]

The anions which are tightly bound by alkylammonium ion micelles[37] are characterized by relatively low hydration energies and high affinities for the tetraalkylammonium ion resin, Dowex 2.[41] They are effective at salting agar out of aqueous solution, and consequently head the lyotropic series.[32] Many related phenomenological series have been identified.[42] Some of this data appears in Table 12, which demonstrates that precisely those anions which are tightly bound to alkylammonium ion micelles are the anions implicated in the micellar stereochemical control of deamination reactions.

The stereochemical salt effects differ from the commonly observed competitive inhibition of micellar catalysis, which is generally due to exclusion from the micelle of a reactive counterion by an impotent counterion.[43] In our work, it seems likely "that the micelles themselves have been altered through strong binding of certain counterions, which, though not incorporated into product, modify product formation occurring within the Stern layers of which they are a part."[34,44]

What is not now clear is the _microscopic_ mechanism(s) by which the cooperative micelle-anion effect is exerted. Is the effect simply a "medium effect"? Do the _tightly_ bound anions so alter the micelles and reduce the aquation of their Stern layers that the deaminations occur in an effectively "more organic solvent"? This would amount to a new way of educing a well-known result from the realm of homogeneous deamination chemistry.[2] Or do the tightly bound anions themselves participate in product formation? Might they shield the back face of the substrate's chiral carbon as nitrogen departs and thus enhance the return/retention path of Scheme II? Current experimentation in our laboratory aims at answering these questions. We also seek to improve our ability to control the stereochemistry of deamination (and related reactions) with micelles.

ASYMMETRIC INDUCTION

The importance of the micellar stereochemical effects reported above lies more in their novelty than in their chemical utility. We have therefore begun to explore the poss-

Table 12

Properties of Anions and the Stereochemical Micellar Control of Deamination

Anion	Control[a]	$\Delta H°_{298}$(Hydration)[b]	Aff. Dowex 2[c]	N[d]
ClO_4^-	+	-57.1	32	11.8
p-$C_7H_7SO_3^-$	+	-----	14	----
BF_4^-	+	-71.2	----	----
I^-	e	-69.7	8-13	12.5
Br^-	-	-79.8	3.4	11.3
NO_3^-	e	-----	3.3	11.6
NO_2^-	f	-----	----	10.1
Cl^-	-	-87.6	1.0	10.0
CH_3COO^-	-	-----	0.17	----
F^-	e	-121.9	0.10	4.8

[a] Ability to influence the stereochemistry of the micellar deamination of 2-aminooctane; see Table 11. [b] In Kcal/mole, see reference 40. [c] Reference 41. [d] The lyotropic number; the relative ability of NaX to salt agar out of water. N_{Cl^-} and $N_{SO_4^=}$ are defined as 10.0 and 2.0, respectively. See reference 32. [e] Not studied. [f] Normally present during deamination. Because Br^- does not assist micellar stereochemical control, NO_2^-, which has a lower N value, probably does not.

ibility of asymmetric induction with micelles composed of chiral units. Perhaps the only literature precedent was the report that the rate of racemization of micelle-solubilized 2-dimethylamino-2'-trimethylammoniumbiphenyl is similar in the presence of either an achiral p-tosylate or a chiral d-10-camphorsulfonate anion.[46]

The absence of chemically significant diastereomeric interactions in the above case may be related to our finding that the deamination of micellized d,l-2-octylammonium ions in the presence of d-10-camphorsulfonate counterions gives only inactive 2-octanol.

Micellar asymmetric induction would appear more likely to succeed if the chiral sites were at or adjacent to the head groups, rather than in the counterions. Indeed, Bunton has found that micelles derived from D(-)-ephedrine selectively catalyze the hydrolysis of D(-)-p-nitrophenyl methoxymandelate, relative to the L enantiomer.[47] The difference, however, was small, apparently less than 10% at the rate constants' maxima. Noting the low stereoselectivity in these experiments, we were not surprised (though still disappointed) by the failure of our own initial, and less quantitative attempts.

Partial hydrolyses of (1) d,l-2-octyl acetate, solubilized by micelles of l-2-octyltrimethylammonium bromide;[11] of (2) d,l-2-octylacetate, solubilized by micelles of l-VI;[48] and of (3) d,l-α-methylbenzyl decanoate, solubilized by micelles of l-VI afforded, in every case, the appropriate racemic alcohol.

$$\underset{VI}{\text{Ph}-\underset{H}{\overset{CH_3}{\underset{|}{\overset{|}{C}}}}-\overset{CH_3}{\underset{CH_3}{\underset{|}{\overset{|}{N}}}}-\underline{n}-C_{16}H_{33}, Br^-}$$

These reactions were carried out in water, at NaOH titers ranging from 0.1-0.2N, and to approximately 25% completion. The alkylammonium ion concentrations were substantially above the cmc ranges. Accelerated hydrolytic rates were observed.

We require more inventive architecture of micellar catalysts and reaction cofactors if highly stereoselective micellar systems are to be produced. Perhaps inverted micelles in non-aqueous solvents will play a role in the solution of this problem,[49] or, it may be that no solution will appear, short of very extensive synthetic efforts to tailor-make micelle monomers.

Acknowledgments

We are grateful to the National Science Foundation and to the National Institutes of Health for their generous support of our research. We also wish to acknowledge the following fellowships: A. P. Sloan Foundation (R.A.M.), Johnson and Johnson Company (C.J.T.), Allied Chemical Company (D.W.R.), and N.D.E.A. (W.L.S.).

REFERENCES AND NOTES

(1) R. Piria, Ann. Chem., 68, 343 (1848).
(2) Reviews include: (a) A. Streitwieser, Jr., J. Org. Chem., 22, 861 (1957); (b) J. H. Ridd, Quart. Rev. (London), 15, 418 (1961); (c) M. C. Whiting, Chem. Brit., 2, 482 (1966); (d) E. H. White and D. J. Woodcock, in "The Chemistry of the Amino Group," S. Patai, Ed., Interscience, New York, N.Y., 1968, pp. 440ff; (e) R. A. Moss, Chem. Eng. News, 49, Number 48, 22 November, 1971, p. 28.
(3) R. A. Moss and D. W. Reger, J. Amer. Chem. Soc., 91, 7539 (1969).
(4) T.W.J. Taylor, J. Chem. Soc., 1099 (1928).
(5) L. P. Hammett, "Physical Organic Chemistry," McGraw-Hill Book Co., New York, N.Y., 1940, p.294.
(6) E. Abel, H. Schmid, and J. Schafranik, Z. Physik. Chem., Bodenstein Festband, 510 (1931).
(7) Ir spectra of the scrubbed and collected nitrogen showed neither NO nor NO_2, and only traces of N_2O.
(8) At pH 4, more than 99% of the aminoalkane is in the protonated, alkylammonium ion form.
(9) R. A. Moss and C. J. Talkowski, Tetrahedron Lett., 703 (1971).
(10) The cmc of 2-OTA$^+$Br$^-$ in 1.5$\underline{M}$ NaBr solution is about 0.15$\underline{M}$,[11] and micelles must be present in Case 5.
(11) R. A. Moss and W. L. Sunshine, J. Org. Chem., 35, 3581 (1970).
(12) Two recent reviews are: (a) E. J. Fendler and J. H. Fendler, Adv. Phys. Org. Chem., 8, 271 (1970); (b) E. H. Cordes and R. B. Dunlap, Accts. Chem. Res., 2, 329 (1969).
(13) Separate experiments demonstrated the linear dependence of k_3 on [H$^+$], demanded by (10).

(14) M. T. Behme and E. H. Cordes, J. Amer. Chem. Soc., 87, 260 (1965).
(15) In terms of a thermodynamic cycle relating free and micellized amine and ammonium ion, the micelle "selects" amine from the bulk phase in preference to ammonium ion, for there is no unfavorable electrostatic interaction in the solubilization of the amine, whereas work must be done against the micellar field in order to solubilize the ammonium ion.
(16) See reference 2c for a statement of this view; also see H. Maskill, R. M. Southam, and M. C. Whiting, Chem. Commun., 496 (1965).
(17) For a review, see D. J. Raber and J. M. Harris, J. Chem. Ed., 49, 60 (1972).
(18) A. Streitwieser, Jr., and W. D. Schaeffer, J. Amer. Chem. Soc., 79, 288 (1957).
(19) K. B. Wiberg, Ph.D. Thesis, Columbia University, New York, N.Y., 1950.
(20) R. A. Moss and S. M. Lane, J. Amer. Chem. Soc., 89, 5655 (1967).
(21) M. Vogel and J. D. Roberts, J. Amer. Chem. Soc., 88, 2262 (1966).
(22) P. Brewster, F. Hiron, E. D. Hughes, C. K. Ingold, and P.A.D.S. Rao, Nature, 166, 179 (1950).
(23) F. G. Mann and H.W.G. Porter, J. Chem. Soc., 456 (1964).
(24) B. Halpern and J. W. Westley, Chem. Commun., 34 (1966).
(25) R. H. Pickard and J. Kenyon, J. Chem. Soc., 45 (1911). See reference 21 for a discussion of configurational correlation of sec-aminoalkanes and sec-alkanols.
(26) R. A. Moss, D. W. Reger, and E. M. Emery, J. Amer. Chem. Soc., 92, 1366 (1970), and references cited there.
(27) L. R. Romsted and E. H. Cordes, J. Amer. Chem. Soc., 90, 4404 (1968); P. Mukerjee and A. Ray, J. Phys. Chem., 70, 2144 (1966).
(28) This point was demonstrated by control experiments.
(29) Only the intermediate values will be considered.
(30) E. W. Anacker and H. M. Gose, J. Phys. Chem., 67, 1713 (1963), and references therein.
(31) Actually less nitrite is present; about 8% of the initial nitrite is protonated at pH 4.
(32) A. Voet, Chem. Rev., 20, 169 (1937).
(33) The tabulated data is omitted for reasons of space, but can be reconstructed from Figure 1.

(34) R. A. Moss and C. J. Talkowski, J. Amer. Chem. Soc., 94, 4767 (1972).
(35) Deaminations in micellar 2-OTA$^+$Br$^-$ solutions (NaNO$_2$/HBr) also failed to afford 2-octanol with altered stereochemistry. However, k_{obs} was augmented.
(36) No doubt further investigation will provide examples of "borderline" anions.
(37) E. W. Anacker and R. D. Geer, J. Coll. and Interface Sci., 35, 441 (1971); R. D. Geer, E. H. Eylar, and E. W. Anacker, J. Phys. Chem., 75, 369 (1971); and references cited there.
(38) In a related study, we observed identical cmc's for racemic and optically active 2-octylammonium ions: "The effect of chirality at the head group... is probably mitigated by the water molecules and gegenions which insulate the head groups from each other in the micelle." See reference 11.
(39) P. Mukerjee, Adv. Coll. and Interface Sci., 1, 241 (1967).
(40) The thermodynamics of anion hydration is discussed by H. F. Halliwell and S. C. Nyburg, Trans. Faraday Soc., 59, 1126 (1963); J. Chem. Soc., 4603 (1960). See, also, S. Subramanian and H. F. Fisher, J. Phys. Chem. 76, 84 (1972).
(41) H. P. Gregor, J. Belle, and R. A. Marcus, J. Amer. Chem. Soc., 77, 2713 (1955).
(42) W. P. Jencks, "Catalysis in Chemistry and Enzymology", McGraw-Hill Book Co., New York, N.Y., 1969, Chapters 7 and 8.
(43) For examples, see: R. B. Dunlap and E. H. Cordes, J. Amer. Chem. Soc., 90, 4395 (1968); C. A. Bunton and L. Robinson, J. Org. Chem., 34, 773, 780 (1969); and references 12a and 12b.
(44) Similar phenomena may be involved in the micelle catalyzed decarboxylation of 6-nitrobenzisoxazole-3-carboxylate ion.[45]
(45) C. A. Bunton, M. J. Minch, and L. Sepulveda, J. Phys. Chem., 75, 2708 (1971); C. A. Bunton, A. Kamego, and M. J. Minch, J. Org. Chem., 37, 1388 (1972).
(46) W. H. Graham and J. E. Leffler, J. Phys. Chem., 63, 1274 (1959).
(47) C. A. Bunton, L. Robinson, and M. F. Stam, Tetrahedron Lett., 121 (1971).
(48) R. A. Moss and W. L. Sunshine, unpublished work.
(49) See, in this regard, E. J. Fendler, J. H. Fendler, R. T. Medary, and V. A. Woods, Chem., Commun., 1497 (1971).

CATALYSIS BY INVERSE MICELLES IN NON-POLAR SOLVENTS

E. J. Fendler, Shuya A. Chang, J. H. Fendler,
R. T. Medary, O. A. El Seoud and V. A. Woods

Department of Chemistry
Texas A & M University
College Station, Texas 77843

The rates of numerous reactions are affected by the presence of micellar surfactants in aqueous solutions.[1-4] Structural similarities between globular proteins and spherical micelles and the analogies between enzymatic and micellar catalysis have prompted recent investigations of micellar systems as possible models for the micro-environment of the active site of enzymes. Although the kinetics for micellar catalysis generally obeys the Michaelis-Menten equation and, in many cases, competitive inhibition has been observed, micelles in aqueous solutions rarely enhance the rates of reactions by factors greater than 10^2 and show relatively limited substrate specificity.[1-3] Micelles, unlike enzymes, are in a dynamic equilibrium with the monomeric surfactant and have comparatively mobile structures in water.[5] Additionally, micelles do not bind the substrate in a rigid configuration with a specific orientation. It appears, therefore, that the aqueous micellar systems investigated to-date provide somewhat poorer models for enzymatic interactions than originally anticipated. Since the active sites of many enzymes are in a relatively hydrophobic environment and since X-ray crystallographic studies have indicated ion pair and hydrogen bonding interactions in polar regions of some proteolytic enzymes,[6,7] model studies in apolar solvents[8] and at interfaces[9] have provided a better understanding of the mechanisms involved. The hydrolysis of p-nitrophenyldodecanoate has recently been examined in hexanol systems containing water and hexadecyltrimethylammonium bromide under conditions where formation of micelles, "reverse" micelles, and liquid crystalline phases have been demonstrated.[10,11] Rate accelerations of ca. 20-fold have been found in the regions where water is solubilized in the polar interior of the reversed micelle. This rate enhancement was analogous to that observed previously for the reaction of p-nitrophenyldodecanoate in aqueous micellar hexadecyltrimethylammonium bromide solution[12] indicating that solubilization of the hydrolyzing agent results in catalytic efficiency similar to that for substrate solubilization.

An alternative, and possibly more informative, approach is to use less polar solvents, such as benzene or cyclohexane, and to limit the surfactant concentrations to those ranges where "reversed" micelle formation predominates. Such systems possess a novel and in many respect unique micro-environment and can provide information with wide applicability to several areas of importance. In particular:

(1) Like the active sites of many enzymes, reversed micelles contain a cavity with polar functional groups capable of binding substrates fairly strongly in specific orientations. Such binding can result in rate enhancements in excess of those expected on the basis of simple partitioning of the substrate between the hydrophilic micellar and bulk hydrophobic phases.

(2) Phospholipids, such as phosphatidyl choline (lecithin), ethanolamine, and serine, also form reversed micelles in non-polar solvents and such micelles have been suggested to be suitable models for cell membranes.[5] Both substrate binding and catalysis in these "naturally occurring" reversed micellar systems can be easily investigated and related to those in structurally similar synthetic surfactants.

(3) It is possible to solubilize a considerable amount of water in the interior of reversed micelles. This reservoir of trapped water can have an ionic strength far in excess of that in aqueous solutions and corresponds to the ionic environment of crystals.[13]

(4) The chemical nature of the cavity can be easily altered to investigate separately and concomittantly the effects of (a) the size of the cavity, (b) its water content, (c) changes in the polarity and structure of the polar

functional groups, and (d) coordinately bonded and "free" metal ions on both substrate interactions and catalysis. Reversed micelles can provide, thereby, a promising model for the micro-environment of enzyme active sites.

(5) Investigations of reactions in these reversed micellar systems, in addition to providing basic and specific information on biologically important catalytic mechanisms, also have an extensive applicability to both large and small scale reactions of industrial importance.

With the above objectives in mind we have initiated structural and kinetic investigations utilizing reversed micelles in non-polar solvents. Additionally we have investigated the micellar properties of these systems using proton magnetic resonance spectroscopy. The ensuing discussion will summarize our present state of understanding.

PROTON MAGNETIC RESONANCE INVESTIGATIONS OF MICELLE FORMATION IN NON-POLAR SOLVENTS

Initial investigations included alkylammonium carboxylate surfactants in cyclohexane, benzene, carbon tetrachloride, chloroform, and dichloroethylene.[14] The ^1H nmr spectra of these surfactants exhibit single weight-averaged resonance frequencies for the protons of the monomeric and aggregated surfactants indicating that the monomer-micelle equilibrium is rapid on the nmr time scale. Chemical shifts for the propionate methyl protons of alkylammonium propionates as a function of the concentration of the surfactant (BAP, HAP, and OAP) in benzene at 33° are given in Figure 1. For each surfactant there is a pronounced break in the chemical shifts at a given concentration. Assuming an idealized situation, neglecting the activity coefficients, the equilibrium between the monomeric, S, and micellar, Sn, surfactant can be written:

$$nS \rightleftharpoons S_n \qquad (1)$$

$$K = \frac{[S_n]}{[S]^n} \qquad (2)$$

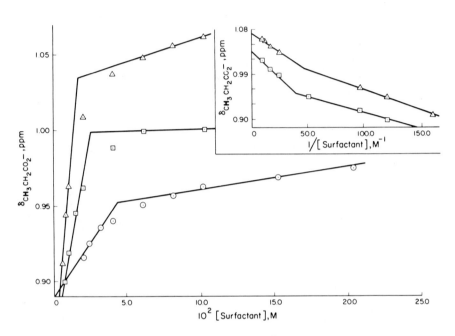

Chemical shifts of the C$\underline{\text{H}}_3$CH$_2$CO$_2^-$ protons as functions of surfactant concentration in benzene at 33°: ⊙, butylammonium propionate (BAP; ▫, hexylammonium propionate (HAP); △, octylammonium propionate (OAP)

where K is the equilibrium constant for the formation of the micelle and n is the aggregation number. Assuming that the concentration of monomers remains essentially constant above the critical micelle concentration, CMC, (equation 3), the

$$n[S_n] = C_D - [S] \qquad (3)$$

observed chemical shift at a given concentration, δ, is described by equation 4,[15] where δ_M and δ_m are the micellar and monomeric chemical shifts, respectively. Plots of the observed chemical shift, δ, against the reciprocal stoichiometric surfactant concentration, $1/C_D$, yielded straight

$$\delta = \delta_M + \frac{CMC}{C_D}(\delta_M - \delta_m) \qquad (4)$$

lines for many aqueous systems with intersections at the CMC's.[15-18] Treatments of the present data, in non-polar benzene according to equation 3 are illustrated in the insert in Figure 1. Using these plots, values for δ_M and δ_m can be calculated. Knowledge of δ_M and δ_m affords the calculation of the concentration of monomeric surfactant, [S], at a given C_D value:

$$[S] = \frac{C_D(\delta_M - \delta)}{\delta_M - \delta_m} \qquad (5)$$

With this information, n and K may be obtained in the following manner: multiplying equation 2 by n and taking the logarithm, equation 6 is obtained:

$$\log(nK) = \log(n[Sn]) - n\log[S] \qquad (6)$$

and substitution of equation 3 and rearrangement leads to:

$$\log(C_D - [S]) = \log nK + n\log[S] \qquad (7)$$

From plots of the left hand side of equation 7 against log [S], n and K can be obtained from the slopes and intercepts, respectively. Treatment of our data according to equation 7 leads to good linear relationships (Figure 2). Additionally, it allows the calculation of the concentrations of the monomer and the micelle at given surfactant concentrations (for example, Figure 3). Apparently, the behavior of monomers and micelles in benzene is analogous to that expected of "normal" micelles in aqueous solutions. Added confidence in the above treatment is gained by extending it to other surfactant protons. Figure 4 illustrates the behavior of octylammonium tetradecanoate protons in benzene as a function of surfactant concentration. Using proton magnetic resonance spectroscopic techniques we have obtained several parameters for alkylammonium carboxylate micelles (Table I). Salient points arising from these results are:

Figure 2

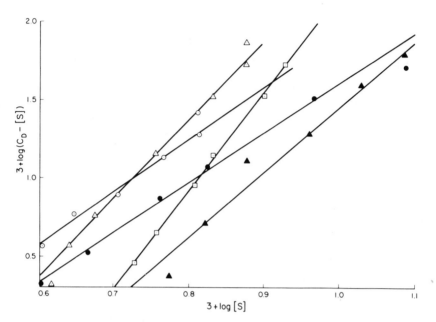

Plot of data according to equation 7. BAP, HAP and OAP in benzene (⊙, ⊡, △) and in carbon tetrachloride (●, ■, ▲), respectively.

(1) Polar carboxylic and ammonium headgroups are in the interior of the aggregate while the hydrocarbon chain is in contact with the non-polar solvent.

(2) The critical micelle concentration of alkylammonium propionates increases with a decreasing number of carbon atoms in the alkyl chain in benzene but remains essentially unchanged in carbon tetrachloride. For octylammonium carboxylates, changes in the chain length of the carboxyl group does not alter the CMC in benzene. Similarly, substitution of halogens in dodecylammonium carboxylates affects the CMC's in benzene insignificantly. Polar solubilizates, however, lower the CMC's appreciably.

(3) Alkylammonium carboxylates form small micelles with aggregation numbers in the range of 3-5.

Figure 3

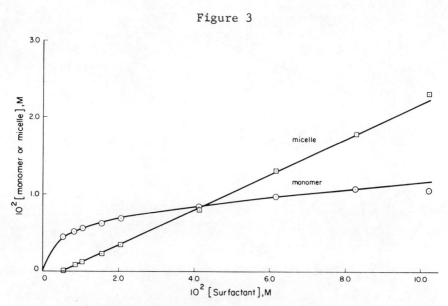

Calculated concentrations of monomeric and micellar HAP in benzene

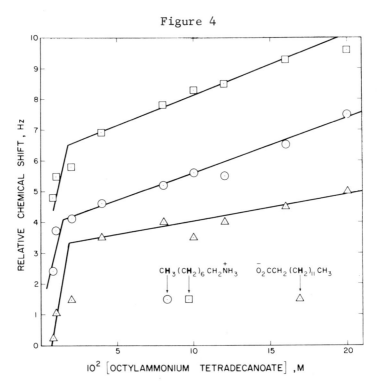

Figure 4

Chemical shifts of octylammonium tetradecanoate protons in benzene at 33° as a function of solute concentration

Table I

Micellar Parameters for Alkylammonium Carboxylates

Surfactant	Benzene			Carbon tetrachloride		
	CMC, M	n	K, M^{1-n}	CMC, M	n	K, M^{1-n}
Butylammonium propionate	$(4.5-5.5)10^{-2}$	4	1×10^4	$(2.3-2.6)10^{-2}$	3	9×10^2
Hexylammonium propionate	$(2.3-3.2)10^{-2}$	7	5×10^{12}	$(2.1-2.4)10^{-2}$	7	7×10^{11}
Octylammonium propionate	$(1.5-1.7)10^{-2}$	5	1×10^8	$(2.6-3.1)10^{-2}$	5	5×10^7
Decylammonium propionate	$(8-10)10^{-3}$			$(2.2-2.7)10^{-2}$	5	1×10^7
Dodecylammonium propionate	$(3-7)10^{-3}$			$(2.1-2.5)10^{-2}$	4	5×10^4
Hexadecyltrimethylammonium propionate	2.4×10^{-3}					
Dodecylammonium 3-bromo-propionate	$(2-3)10^{-3}$					
Dodecylammonium 3-iodo-propionate	$(2-3)10^{-3}$					
Octylammonium butyrate	$(2.7-3.6)10^{-2}$					
Octylammonium hexanoate	$(2-6)10^{-2}$	3				
Octylammonium nonanoate	$(1-6)10^{-2}$	3				
Octylammonium tetradecanoate	$(1.9-2.2)10^{-2}$					
Octylammonium benzoate	$(5-6)10^{-2}$	3				

INTERACTIONS IN AND CATALYSIS BY REVERSED MICELLES IN NON-POLAR SOLVENTS

Up to the present time we have observed pronounced micellar catalysis of the mutarotation rate of 2,3,4,6-tetramethyl-α-D-glucose and the rate of decomposition of 2,4,6-trinitrocyclohexadienylide ion in benzene and cyclohexane.[19,20]

Changes in the observed mutarotation rate constant, k_ψ ($k_\psi = k_{forward} + k_{reverse}$) for 2,3,4,6-tetramethyl-α-D-glucose in benzene and in cyclohexane as a function of dodecylammonium propionate (DAP) concentration are illustrated in Figure 5. As in aqueous micellar catalyses,[1-5] the rate constant-concentration profile for each surfactant exhibits a sigmoidal dependence followed by a plateau. In the plateau region the rate constants for the mutarotation of 2,3,4,6-tetramethyl-α-D-glucose in benzene in the presence of DAP, DABz and DAB are factors of 380, 457 and 688 greater than those in the pure solvent. The rate enhancement by DAP in cyclohexane is a factor of 863. The magnitude of these rate enhancements is considerably greater than those generally observed in aqueous micellar systems.[1-3] It is also noteworthy that micellar effects in cyclohexane are considerably more pronounced than those in benzene. Our qualitative observation that the solubility of 2,3,4,6-tetramethyl-α-D-glucose is less in cyclohexane than in benzene suggests a more favorable partitioning between the polar micellar phase and the bulk apolar solvent in the former than in the latter.

The rate acceleration of the mutarotation of 2,3,4,6-tetramethyl-α-D-glucose in benzene is not due to the presence of small amounts of water in the benzene, since the rate constant in "wet" benzene is only slightly greater in dry benzene. More significantly, the k_ψ value in water at pH 5.43 (34.5×10^{-5} sec^{-1}) is factors of 15, 18 and 28 smaller than that in benzene in the presence of micellar DAP, DABz and DAB, respectively. Catalysis of the mutarotation of 2,3,4,6-tetramethyl-α-D-glucose in benzene by DAP, DABz and DAB at the beginning of the plateau is, in fact, factors of 210, 260 and 370 greater than that by hydronium ions in water. In cyclohexane the same factor is 3000. Using equations 8 and 9

CATALYSIS BY INVERSE MICELLES IN NON-POLAR SOLVENTS

Figure 5

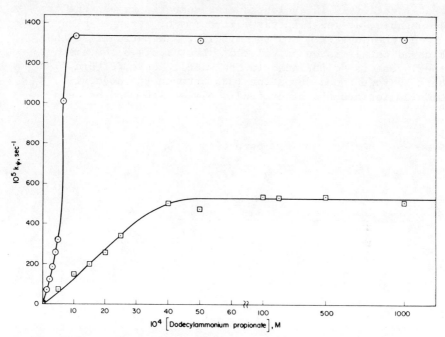

k_ψ values for the mutarotation of 2,3,4,6-tetramethyl-α-D-glucose as a function of DAP concentration

$$\frac{1}{k_o - k_\psi} = \frac{1}{k_o - k_m} + \frac{1}{k_o - k_m}\left(\frac{1}{C_D - CMC}\right)\frac{N}{K} \quad (8)$$

$$\frac{k_\psi - k_o}{k_m - k_\psi} = \frac{K}{N}(C_D - CMC) \quad (9)$$

where k_o and k_m are the rate constants in the bulk and micellar phases, respectively, C_D is the stoichiometric surfactant concentration, K is the micelle-substrate equilibrium constant and N is the aggregation number, values for K/N have been calculated from good linear plots for DAP

(Figure 6) to be 688 in benzene and 863 in cyclohexane. The greater catalysis in cyclohexane as compared to benzene is accountable, at least in part, in terms of stronger binding between the substrate and the micelle. To rationalize the observed catalysis we postulated[19] that 2,3,4,6-tetramethyl-α-D-glucose is solubilized in the micelle core (Figure 7) where hydrogen bond formation both between the dodecylammonium ion and the heterocyclic oxygen atom and between

Figure 6

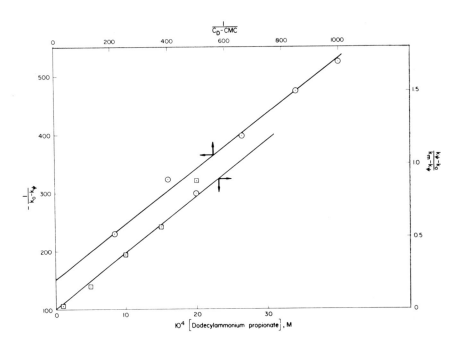

Micelle-substrate association (binding) constant plots for 2,3,4,6-tetramethyl-α-D-glucose (1.7 x 10^{-2} M) and DAP in benzene according to equation 8 (⊙) and equation 9 (⊡).

Figure 7

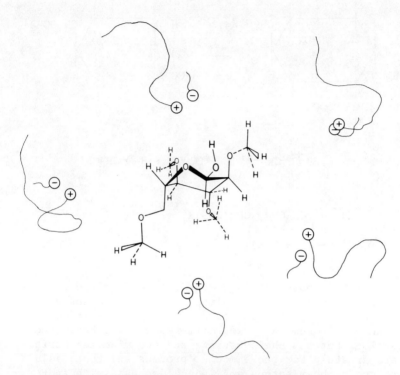

Schematic representation of 2,3,4,6-tetra-methyl-α-D-glucose solubilized in reversed alkylammonium carboxylate micelles

the 1-hydroxyl group and the carboxylate ion of the surfactant can facilite ring opening:

$$\text{[structure of tetramethyl-}\alpha\text{-D-glucose with RCOO}^-\text{ and H}_3\text{N}^+\text{-CH}_2\text{-R}'\text{]} \rightleftharpoons \text{[products]} + \text{RCOOH} + \text{H}_2\text{N-CH}_2\text{-R}'$$

R = CH_3CH_2, $CH_3(CH_2)_2$, or C_6H_5

R' = $CH_3(CH_2)_{10}$

Support for this postulate has been obtained in the pmr investigation of the chemical shifts of the surfactant protons as a function of 2,3,4,6-tetramethyl-α-D-glucose concentration, since the chemical shift differences, and hence the interaction, between the substrate and the micellar surfactant are greatest for the ammonium protons and those adjacent to the carboxyl group and indeed decrease with increasing separation from the charged atoms of the head groups.[19]

In a similar investigation we examined the rate of decomposition of Meisenheimer, or σ-, complexes.[20] These complexes are intermediates in nucleophilic aromatic substitutions and their isolation as alkali metal cyclohexadienylides allows the independent investigation of the kinetics of their unimolecular decomposition.[21] We have studied the decomposition of sodium 1,1-dimethoxy-2,4,6-trinitrocyclohexadienylide in benzene in the presence of micellar surfactants. In micellar dodecylammonium benzoate in benzene the rate constant for the decomposition of the complex, k_ψ, is 62,900-fold greater than in pure benzene, and, more significantly, it is greater by a factor of 1880 than that in pure water. This latter observation clearly indicates that rate enhancement <u>is not</u> the result of simple partitioning of the complex between the hydrophilic micellar interior and the hydrophobic bulk solvent. It is likely that the

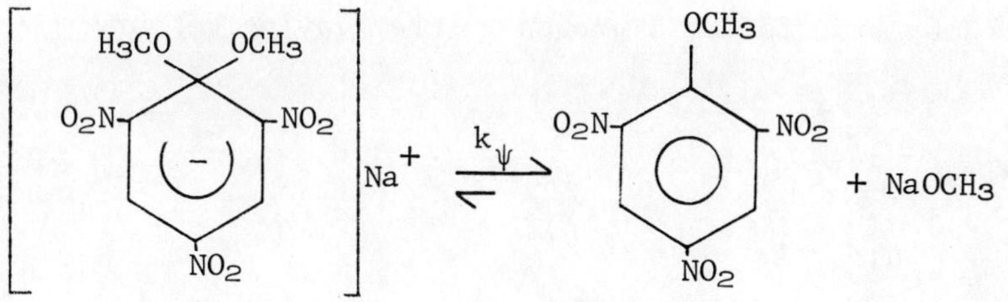

CATALYSIS BY INVERSE MICELLES IN NON-POLAR SOLVENTS

$$\left[\begin{array}{c} H_3CO \quad OCH_3 \\ O_2N \underset{NO_2}{\underset{|}{\bigcirc}} NO_2 \\ \end{array} \right] Na^+ \underset{}{\overset{k_\psi}{\rightleftharpoons}} \begin{array}{c} OCH_3 \\ O_2N \underset{NO_2}{\underset{|}{\bigcirc}} NO_2 \end{array} + NaOCH_3$$

Meisenheimer complex can be oriented in the micellar cavity (Figure 8) where breaking of the carbon-oxygen bond is

Figure 8

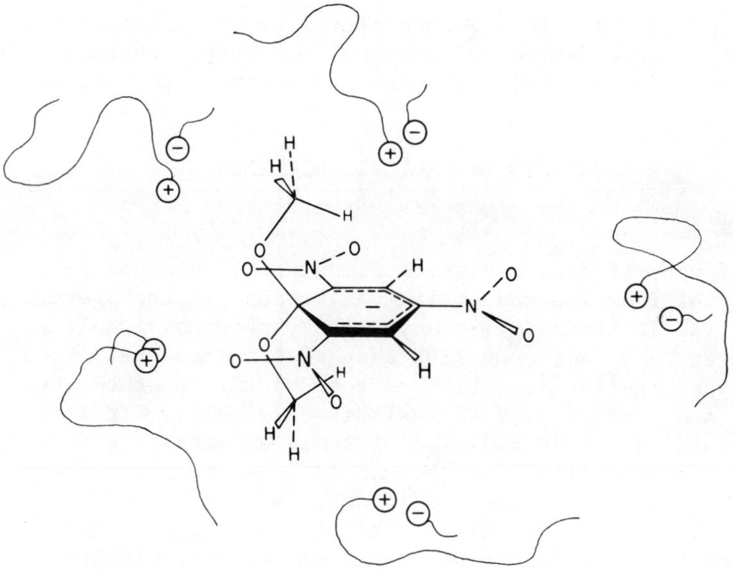

Schematic representation of 2,4,6-trinitro-cyclohexadienylide solubilized in reversed alkylammonium carboxylate micelles

assisted by transfer of a proton to the leaving methoxyl group:

$$R-C\overset{O}{\underset{O}{\rightleftharpoons}} \quad H-\underset{H}{\overset{H}{N}}-CH_2-R'$$

[Meisenheimer complex with H$_3$CO, OCH$_3$, O$_2$N, NO$_2$, NO$_2$ substituents, Na$^+$] ⇌ [trinitroanisole with OCH$_3$, O$_2$N, NO$_2$, NO$_2$] + CH$_3$OH + RCO$_2^-$Na$^+$ + H$_2$N-CH$_2$-R'

Lack of catalysis of the decomposition of the complex by micellar hexadecyltrimethylammonium propionate, where proton donation is not possible, is consistent with this postulate.

CATALYSIS BY LECITHIN MICELLES IN NON-POLAR SOLVENTS

The rate of decomposition of sodium 1,1-dimethoxy-2,4,6-trinitrocyclohexadienylide in benzene in the presence of micellar lecithin (β,γ-diacylglycerylphosphorylcholine) is greater by a factor of 2100 than that in benzene. Saturation type kinetics are observed both with respect to the complex and with respect to lecithin (Figure 9). It is even more significant that addition of small amounts of water, known to be solubilized at the hydrophilic interior of the lecithin micelle,[22] causes additional rate enhancement at a given lecithin concentration. The profile for this rate enhancement is also sigmoidal; maximum catalysis (at 1.2×10^{-1}% water) occurs when the micellar interior presumably is saturated with water. At this point the overall rate enhancement is ca. 16,000-fold, approaching a value observed in the case of catalysis by micellar surfactants in benzene. The origin of catalysis by lecithin micelles in non-polar benzene can be discussed analogously to that postulated for synthetic reversed micelles. The polar

Figure 9

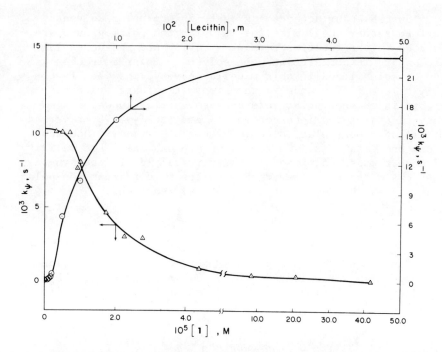

Rate constants for the decomposition of 1,1-dimethoxy-2,4,6-trinitrocyclohexadienylide ion (<u>1</u>), k_ψ, in benzene at 24.6° as a function of lecithin concentration at [<u>1</u>] = 5.05 x 10^{-6} M (⊙); and as a function of the concentration of (<u>1</u>) at [lecithin] = 50.0 x 10^{-4} M (△).

substrate is held in the hydrophilic micellar cavity where transfer of a proton is possible. We are currently examining the effects of micellar lecithin and other phospholipids on the rates of several reactions in non-polar media.

CONCLUSION

Suitable surfactants aggregate in non-polar solvents to form rather small micelles in which the polar groups are located in the interior forming a cavity and the non-polar groups are in contact with the solvent. Polar molecules can be solubilized in this micellar cavity, and when favorably oriented they undergo reactions at a faster rate than either in the non-polar reference solvent or in water. Many facets of the interactions in and catalysis by reversed micelles in non-polar solvents resemble enzymatic interactions and catalysis. Currently we are examining the validity of such a model using several structurally different synthetic and "natural" reversed micelles in different non-polar solvents as the media for a variety of reactions.

ACKNOWLEDGMENTS

Support of the work by the Robert A. Welch Foundation is gratefully acknowledged. E. J. F. is a Career Development Awardee of the National Institutes of Health, U. S. Public Health Service.

REFERENCES

1. E. H. Cordes and R. B. Dunlap, Accounts Chem. Res., 2, 329 (1969).

2. E. J. Fendler and J. H. Fendler, Advan. Phys. Org. Chem., 8, 271 (1970).

3. T. C. Bruice, in "The Enzymes," Vol. 2, 3rd ed., Academic Press, New York, N.Y., 1970, p. 217.

4. H. Morawetz, Advan. Cat., 20, 341 (1969); Accounts Chem. Res., 3, 354 (1970).

5. P. H. Elworthy, A. T. Florence, and C. B. Macfarlane, "Solubilization by Surface Active Agents," Chapman and Hall, London, 1968.

6. P. B. Sigler, D. M. Blow, B. W. Matthews, and R. Henderson, J. Mol. Biol., 35, 143 (1968).

7. T. A. Steitz, R. Henderson, and D. M. Blow, J. Mol. Biol., 46, 337 (1969).

8. F. M. Menger, J. Amer. Chem. Soc., 88, 3081 (1966); R. L. Snell, W. Kwok, and Y. Kim, ibid., 89, 6728 (1967).

9. F. M. Menger, ibid., 92, 5965 (1970).

10. S. Friberg and S. I. Ahmad, J. Phys. Chem., 75, 2001 (1971).

11. S. I. Ahmad and S. Friberg, J. Amer. Chem. Soc., 94, 5196 (1972).

12. L. R. Romsted and E. H. Cordes, J. Amer. Chem. Soc., 90, 4404 (1968).

13. F. M. Fowkes, in "Solvent Properties of Surfactant Solutions," K. Shinoda, Ed., Marcel Dekker, New York, N.Y., 1967, p. 67.

14. J. H. Fendler, E. J. Fendler, R. T. Medary, and O. A. El Seoud, J.C.S. Faraday I, submitted.

15. N. Muller and R. H. Birkhahn, J. Phys. Chem., 71, 957 (1967).

16. N. Muller and R. H. Birkhahn, ibid., 72, 583 (1968).

17. R. Haque, ibid., 72, 3056 (1968).

18. R. E. Bailey and G. H. Cady, ibid., 73, 1612 (1969).

19. E. J. Fendler, J. H. Fendler, R. T. Medary, and V. A. Woods, Chem. Comm., 1497 (1971); J. H. Fendler, E. J. Fendler, R. T. Medary and V. A. Woods, J. Amer. Chem. Soc., 94, 0000 (1972).

20. J. H. Fendler, J.C.S. Chem. Comm., 269 (1972).

21. J. H. Fendler, E. J. Fendler, and M. V. Merritt, J. Org. Chem., 36, 2172 (1971); L. M. Casilio, E. J. Fendler, and J. H. Fendler, J. Chem. Soc. (B), 1377 (1971); and references cited therein.

22. W. V. Walter and R. G. Hayes, Biochim. Biophys. Acta, 249, 528 (1971).

AUTHOR INDEX

Abel, E., 100
Adams, G. E., 56
Ahmad, S. I., 76, 128
Alam, A., 35
Albrizzio, J., 47
Alder, R. W., 73
Anacker, E. W., 77, 113, 114, 120, 121
Anbar, M., 54, 59
Anderson, E., 30
Archila, J., 47
Armas, A., 35, 41, 42
Arrington, P. A., 31, 32
Atherton, N. M., 5
Auburn, J. J., 6, 7

Bailey, R. E., 131
Baker, R., 73
Bansal, K. M., 59, 61, 65
Baumrucker, J., 35, 41, 42, 83
Behme, M. T. A., 34, 37, 41, 104
Benkovic, S. J., 76
Belle, J., 121
Bellorin, C., 11, 14, 15, 16
Benjamin, L., 11, 14, 15, 16
Bennion, B. C., 6
Bernstein, H. J., 3
Birkhahn, R. H., 3, 4, 6, 10, 130, 131
Bishop, W., 30
Blow, D. M., 127
Boag, J. W., 56
Bogan, G., 59, 65
Bordwell, F. G., 93
Braams, R., 55
Brewster, P., 108, 110

Bruice, T. C., 76, 127, 136
Brod, L., 30, 42
Brown, H. C., 90
Brown, J. M., 73
Buist, G. J., 78, 94
Bull, H. G., 42
Bunton, C. A., 30, 37, 73, 74, 76, 77, 78, 79, 81, 82, 92, 93

Cadet, J., 63
Cady, G. H., 131
Calzadilla, M., 35, 41, 42, 83
Casilio, L. M., 140
Centeno, M., 41, 83
Chen, W. W., 4, 10, 16
Child, W. C., Jr., 18
Claussen, W. F., 18
Clemente, H., 35, 41, 42
Clifford, J., 11, 12, 17, 18, 19
Clouse, A., 31, 32
Cohen, T., 77
Corkill, J. M., 11, 14, 16
Cordes, E. H., 1, 25, 26, 30, 31, 32, 33, 34, 35, 36, 37, 38, 39, 40, 41, 42, 47, 49, 56, 74, 76, 83, 103, 104, 109, 120, 121, 127, 128, 136
Coronel, J., 35, 41, 42
Courchene, W. L., 8, 11
Cox, J. R., 73, 74
Cram, D. J., 90
Creazzola, A., 35, 41, 42
Cuenca, A., 35, 41, 42
Current, J., 56

Dahlquist, P. W., 26
Davenport, G., 6, 8
Day, C. L., 10, 60
DeWolfe, R. A., 30
Dianoux, A. C., 8, 9
Doddrell, D., 31, 32
Dunham, D., 41, 83
Dunlap, R. B., 1, 25, 31, 32, 33, 34, 35, 36, 38, 39, 40, 42, 74, 76, 103, 104, 109, 120, 121, 127, 136
Duynstee, E. F. J., 47

Ebert, M., 55
El Seoud, O. A., 129
Elworthy, P. H., 76, 127, 128, 136
Emery, E. M., 109
Eriksson, J. C., 4, 10, 13, 63
Eylar, E. H., 120, 121
Eyring, E. M., 6, 7

Farber, S. J., 74
Fendler, E. J., 1, 10, 11, 25, 37, 53, 55, 56, 58, 59, 60, 61, 62, 63, 65, 73, 74, 76, 77, 78, 79, 103, 120, 123, 127, 129, 136, 138, 140
Fendler, J. H., 1, 10, 11, 25, 53, 55, 56, 58, 59, 60, 61, 62, 63, 65, 73, 74, 76, 85, 92, 103, 120, 123, 127, 129, 136, 138, 140
Fife, T. H., 30, 42
Fisher, H. F., 120
Florence, A. T., 76, 127, 128, 136
Fowkes, F. M., 128
Francis, J., 35, 41, 42
Friberg, S., 76, 128
Fullington, J., 34, 37, 41

Geer, R. D., 120, 121
Ghanim, G. A., 35, 42
Ghose, H. M., 77, 113, 114, 120
Gillberg, G., 4, 10, 13, 60
Gitler, C., 8, 9, 25, 26
Glew, D. N., 18
Goodman, J. F., 11, 14, 16
Gordon, J. E., 10
Graber, E., 5
Graham, W. H., 122
Greenstock, C. L., 56
Gregor, H. P., 121
Griffith, J. H., 8
Grossweiner, L. I., 55
Grunwald, E., 47

Hahn, B. -S., 63
Halliwell, H. F., 120
Halpern, B., 109
Hammett, L. P., 73, 100
Hamori, E., 6, 8
Haque, R., 131
Harris, J. M., 106
Hart, E. J., 59
Hayes, R. G., 142
Henderson, R., 127
Herries, D. G., 30
Hertz, H. G., 19
Hiron, F., 108, 110
Holdren, G. R., 57
Hughes, E. D., 108, 110
Hunt, J. W., 56

Ingold, C. K., 73, 108, 110
Inoue, H., 4
Isemura, T., 76

Jeffrey, G. A., 18
Jencks, W. P., 76, 121
Johnson, T. W., 4, 10, 16

AUTHOR INDEX

Kamego, A., 82, 93
Kauzmann, W., 14, 16
Keith, A. D., 8
Kenyon, J., 109
Kim, C. J., 90
Kim, Y., 127
Kirby, A. J., 73
Knipe, A. C., 93
Koehler, K., 42
Kresheck, G. C., 6, 8
Kuo, I., 59, 68
Kurz, J. L., 76
Kwok, W., 127

Lane, S. M., 107, 108, 110
Lang, J., 5, 6, 7
Leffler, J. E., 122
Lehrmann, G., 41, 83
Lindenbaum, S., 19
Lindquist, P., 41, 83

MacFarlane, C. B., 76, 127, 128, 136
Mak, H. D., 18
Malpica, A., 35, 41, 42
Mann, F. G., 109
Marcus, R. A., 121
Maskill, H., 125
Matthews, B. W., 127
Medary, R. T., 123, 129, 136, 138, 140
Menger, F. M., 77, 127
Merritt, M. V., 140
Michael, B. D., 56
Minch, M. J., 81, 82, 93, 122
Miura, M., 5
Moelwyn-Highes, E. A., 18
Morawetz, H., 74, 76, 127, 136,
Moss, R. A., 99, 100, 101, 103, 106, 107, 108, 109, 110, 117, 121, 123
Mukerjee, P., 3, 8, 13, 14, 16, 26, 109, 120

Muller, N., 3, 4, 7, 10, 16, 17, 130, 131
Mysels, K. J., 3

Nakagawa, T., 4, 76
Narten, A. H., 19
Nemethy, G., 14, 15, 16, 19
Neta, P., 54, 57
Ng, M., 56
Noel, R., 34, 37, 41
Nyburg, S. C., 120

Oguri, H., 5
Ortiz, J. J., 42

Patterson, L. K., 10, 59, 61, 65, 85, 92
Pellerin, J. H., 4, 10, 16, 17
Pethica, B. A., 11, 12, 17, 18, 19
Pickard, R. H., 109
Piria, R., 99
Platko, F. E., 4
Pletcher, T. C., 42
Poland, D. C., 14
Polglase, M. F., 18
Pople, J. A., 3
Porter, H. W. G., 109
Portnoy, C. E., 77
Price, M., 41, 83

Quintero, D., 35, 41, 42

Raber, D. J., 106
Raftery, M. A., 26
Ramsay, O. B., 73, 74
Rand-Meir, T., 26
Rao, P. A. D. S., 108, 110
Rath, N. S., 18

Ray, A., 26, 109
Reger, D. W., 99, 108, 109
Rehfeld, S. J., 10, 61
Richards, F. M., 30
Ridd, J. H., 99, 100, 106, 109, 121
Roberts, J. D., 108
Robertson, J. C., 10
Robinson, L., 75, 76, 78, 92
Rodulfo, T., 47
Romero, R., 35, 41, 42
Romsted, L. R., 49, 76, 109, 128

Salazar, J., 35, 41, 42
Sanchez, N., 35, 41, 42
Schaeffer, W. D., 106, 107
Schafranik, J., 100
Scheraga, H. A., 6, 8, 14, 15, 16, 19
Schmid, H., 100
Schneider, W. G., 3
Schott, H., 76
Schuler, R. H., 57
Sears, B., 41, 83
Sepulveda, L., 37, 77, 78, 79
Shinitzky, M., 8, 9
Shinoda, K., 76, 128
Sigler, P. B., 127
Snell, R. L., 127
Southam, R. M., 125
Spalthoff, W., 19
Stam, M. F., 78, 94, 123
Steitz, T. A., 127
Strach, S. J., 5
Streitwieser, A., 99, 100, 106, 107, 109, 121
Subramanian, S., 120
Sunshine, W. L., 103, 123

Talkowski, C. J., 101, 117, 121
Tate, J. R., 14

Tamamushi, B., 76
Taylor, T. W. J., 100
Teoule, R., 58, 63
Thorne, R. L., 10
Tori, K., 4

Varvoglis, A. G., 73
Vassilliades, T., 77
Voet, A., 114, 121
Vogel, M., 108
Von Bergen, R., 35, 41, 42

Waggoner, A. S., 8
Walker, T., 11
Walter, W. V., 142
Wang, S. Y., 63
Ward, J. F., 59, 65
Weber, G., 8, 9
Westley, J. W., 109
White, E. H., 99, 100, 106, 109, 121
Whiting, M. C., 99, 100, 105, 106, 109, 121
Wiberg, K. B., 107
Woodcock, D. J., 99, 100, 106, 109, 121
Woods, V. A., 123

Yang, K. -U., 37, 77, 78, 79
Yasunaga, T., 5

Zana, R., 5

SUBJECT INDEX

Acetals, mechanism of hydrolysis, 28
n-Alkanols, 15
Alkylammonium carboxylates, 129, 135
Alkylammonium propionate, 129
Alkyl carboxylates, 5
Alkyldiazonium ion, 105
sec-Alkyldiazotic acids, 105
Alkyldisulfonates, 34
n-Alkyl hexaoxyethylene glycol monoethers, 15
n-Alkyltrimethylammonium ions, 47, 48, 102
2-Aminoalkanes, 100
1-Aminobutane, 106, 107
2-Aminobutane, 100, 103, 107, 118
d,ℓ-2-Aminodecane, 117
2-Aminohexane, 100, 109
2-Aminoheptane, 102, 103, 109
2-Aminooctane, 108, 110, 112, 116, 117, 118, 119, 122
Asymmetric induction, 121

Barbituric acid, 68
Benzaldehyde diethyl acetals, 35, 42
Benzene, 4, 10, 13, 61, 81, 85, 92, 128, 129, 131, 136, 138, 140
Benzophenone dimethyl ketal, 35
Benzotrifluoride, 10
Brilliant green, 47
3-Bromo-3-phenyl-propionate, 93
Butane-1-diazotic acid, 106
Butane-2-diazotic acid, 107
2-Butanol, 107
n-Butanol, 45
2-Butylammonium ion, 118
Butylammonium propionate, 130, 135

d-10-Camphorsulfonate, 115, 118, 122, 123
Carbonium ions, 25 ff
Carbon tetrachloride, 129

Cetylpyridinium chloride, 85
Cetyltrimethylammonium bromide, 12, 34, 59, 60, 61, 63,
  74, 78, 79, 80, 81, 84, 85, 86, 87, 91, 92, 93, 94, 104
Cetyltrimethylammonium chloride, 59
Chloroform, 129
Clathrate hydrates, 18
Critical micelle concentration, 2, 13, 56
    for alkylammonium carboxylates, 135
Crystal violet, 46, 47, 48, 49
2-Cyano-2-phenylacetate, 82
Cyclohexane, 128, 129, 136, 138
α-Cyclopropylethylamine, 107

2-Decylammonium ion, 117
Decylammonium propionate, 135
Diazonium ions, 100
Dichloroethylene, 129
2-Dimethylamino-2'-trimethylammoniumbiphenyl, 122
Dimethyldodecylphosphine oxide, 45
2,4-Dinitrophenyl phosphate, 78, 79, 94
2,6-Dinitrophenyl phosphate, 79
Dioxane, 81
Disodium sulfoalkylsulfates, 34
Disodium-2-sulfo-2-butyltetradecanoate, 40
Disodium-2-sulfoethyl α-sulfopalmitate, 40
Disodium-2-sulfoethyl α-sulfostearate, 40
Disodium-2-sulfo-2-methyloctodecanoate, 39, 40
Disodium-2-sulfooctadecyl sulfate, 40
Di-t-butyl nitroxide, 5-6
Dodecylammonium benzoate, 140
Dodecylammonium 3-bromo-propionate, 135
Dodecylammonium chloride, 6
Dodecylammonium 3-iodopropionate, 135
Dodecylammonium propionate, 135, 136
Dodecyldimethylammoniumacetate, 34
Dodecyldimethylammoniumpropanesulfonate, 34, 35
Dodecyldimethylphosphine oxide, 34, 35
Dodecyl pyridinium iodide, 6
Dodecyl sodium sulfonate, 6
Dodecyl sulfate, 19

Electron spin resonance, 5
D(-)Ephedrine, 123

SUBJECT INDEX

Equilibrium constants, for interaction of micelles with organic molecules, 31, 62, 77-78, 112, 129
ESR spectroscopy, 8
Esterone, 81, 95
Ethanol, 45
Exchange broadening, 3, 4

Fluorescence depolarization, 8
p-Fluorobenzaldehyde diethyl acetal, 31
Fluoroborate, 115, 118
Formylpyruvylurea, 58, 65, 68

Gamma rays, 53
General acid catalysis, 30
Gouy-Chapman double layer, 11

Heparin, 55
n-Heptanol, 45
Hexadecylsulfates, 34
Hexadecyltrimethylammonium bromide, 49, 56, 127, 128
Hexadecyltrimethylammonium propionate, 135
Hexylammonium propionate, 130, 135
Hydrated electron, 53
5-Hydroperoxy-6-hydroxydihydrothymine, 64
trans-5-Hydroperoxy-6-hydroxydihydrothymine, 58
5-Hydroxy-6-hydroperoxydihydrothymine, 58
5-Hydroxy-5-methylbarbituric acid, 65
Hydrophobic bonding, 16
Hydroxyl radical, 55
5-Hydroxy-5-methyl barbituric acid, 58, 64

Igepal CO-730, 56, 59, 60, 63, 67
Iminium ions, 104
Isopropylbenzene, 89

Ketals, mechanism of hydrolysis, 28

Lecithin, 83, 128, 143
Lysolecithin, 83
Lysozyme, 25

Malachite green, 47
Meisenheimer complexes, 140, 141
2-Methylanthracene, 9
$\underline{d}$-ℓ-α-Methylbenzyl decanoate, 123
Methylcyclohexane, 85
Methylene blue, 55
Methyl orthoacetate, 41
Methyl orthobenzoate, 30, 31, 33, 34, 36, 37, 38, 39, 40,
   41, 42, 43, 44, 45, 46, 76
Methyl ortho-$\underline{p}$-fluorobenzoate, 31
Methyl orthovalerate, 41
Micelles
    kinetics of formation, 2ff
    dissociation of, 2ff
    dissolution of, 2ff
    counterion dissociation, 2, 5
    guest dissociation, 2, 5
    location of water, 9ff
    interaction with solubilizates, 9ff
    partial molal volume, 11
    Stern layer, 11, 26, 120
    Gouy-Chapman double layer, 11
    thermodynamics of formation, 13ff
    as catalysts for carbonium ion reactions, 25ff, 100ff
    surface characteristics, 26-28
    as catalysts for hydrolytic reactions, 28ff
    in radiation-induced reactions, 55ff
    as catalysts for phosphate ester hydrolysis, 75ff
    as catalysts for decarboxylation, 79ff
    as catalysts for nitrous acid deaminations, 99ff
    inverse, as catalysts, 127ff
    of alkylammonium carboxylates, 131-135
    as catalysts for mutarotation, 136-141
    as catalysts for Meisenheimer complex decomposition,
       140-143
    of lecithin, 142-143
Micellization, heat of, 14

Nitrobenzene, 89
6-Nitrobenzisoxazole-3-carboxylate ion, 79, 80, 81, 83, 84,
   91, 92, 94
$\underline{p}$-Nitrophenyldodecanoate, 127, 128
D(-)-$\underline{p}$-Nitrophenyl methoxymandelate, 123
Nitrous acid, 104
Nitrous acid deamination, mechanism, 99

SUBJECT INDEX

Nitrous anhydride, 104, 105
Nitrous oxide, 55
Nuclear magnetic resonance spectroscopy
    fluorine, 4, 8, 10, 16, 31
    proton, 11, 12, 17, 61, 86-89, 129-135
    shielding parameters, 32

Octadecyldimethylammonium bromide, 35
2-Octanol, 108, 109, 116, 117, 118
d-ℓ-2-Octylacetate, 123
2-Octyl acetyllactates, 109
Octylammonium benzoate, 135
Octylammonium butyrate, 135
Octylammonium hexanoate, 135
2-Octylammonium ion, 114, 115, 118
d-ℓ-2-Octylammonium ion, 123
Octylammonium nonanoate, 135
Octylammonium propionate, 130, 135
Octylammonium tetradecanoate, 131, 134, 135
2-Octyl mesylate, 106
Octylphenyl polyoxyethylene ether, 6
2-Octyltrimethylammonium ion, 103, 113
ℓ-2-Octyltrimethylammonium bromide, 123
Oleic acid, 9
Ortho esters, mechanism of hydrolysis, 28
Oxyethylenesulfates, 34

Partial molal volume, 11
Perylene, 9
Phenol, 92, 94
Phenoxide ion, 92
α-Phenylethylamine, 107
Phsophate ester hydrolysis, 73
Phosphoryl fluoride, 74
Polyoxyethylene(15)nonylphenol, 56
Potassium dodecanoate, 5
Potassium thiocyanate, 82
Pulse radiolysis, 56
Pyruvyl formyl urea, 64

Radiation induced reactions, 53

Salt effects
    on micelle-catalyzed carbonium reactions, 42-44
    on micelle-catalyzed fading of crystal violet, 49
    on micellar catalysis, 73ff
    on micelle-catalyzed ester hydrolysis, 76
    on phosphate ester hydrolysis, 76-79
    on decarboxylation, 79ff
    on nitrous acid deamination, 113ff
Sodium alkyl sulfates, 12, 18, 34, 39, 41
Sodium arenesulfonates, 84
Sodium benzene sulfonate, 84, 88
Sodium cholate, 81, 82
Sodium decylsulfonate, 35
Sodium 1,1-dimethoxy-2,4,6-trinitrocyclohexadienylide, 140, 142
Sodium-2-dodecylbenzenesulfonate, 34
Sodium dodecyl sulfate, 4, 5, 6, 30, 31, 32, 33, 34, 35, 36, 37, 38, 41, 42, 43, 44, 45, 56, 59, 60, 61, 62, 63, 74
Sodium dodecylsulfonate, 41
Sodium-2-hexadecyloxy-1-methylsulfate, 35
Sodium-2-hexadecyloxysulfate, 35
Sodium hexadecyl sulfates, 38
Sodium n-alkyl sulfates, 17
Sodium β-naphthalene sulfonate, 81, 92
Sodium nitrite, 100
Sodium-p-t-butylbenzene sulfonate, 82
Sodium tosylate, 81, 83, 84, 86, 87, 92
Stern layer, 11, 104, 109, 120
Stereochemistry, of nitrous acid deaminations, 105ff
2-(p-Substitutedphenoxy)-tetrahydropyrans, 35
Sulfoalkylcarboxylates, 34
α-Sulfoalkyl esters, 34

Testosterone, 81, 94
2,3,4,6-tetramethyl-α-D-glucose, 136, 138, 139, 140
Thermodynamics of micellization, 13
Thiocyanate ion, 57
Thymine, 58, 63, 65, 67, 68
Thymine dimer, 58, 64, 65, 68
cis-Thymine glycol, 58, 64, 68
trans-Thymine glycol, 58, 64, 68
p-Toluenesulfonate, 115
Trimethylacetate, 82
2,4,6-Trinitrocyclohexadienylide ion, 136, 141

SUBJECT INDEX

Tri-p-anisyl methyl cation, 90

van der Waals interactions, 19
Vinyl phosphate, 76

1250
2/9 0 55